Year 2 Maths Workbook

Addition and Subtraction Practice Book for 6 - 7 Year Olds

Jungle Publishing

GW00392625

Introduction

This book helps kids master the timeless skill of mental arithmetic through addition and subtraction exercises.

It focuses on Year 2 material in line with the National Curriculum and also includes more difficult sums to help pupils get ahead. It can be used, too, as a refresher for all pupils in Lower Key Stage 2.

The book is divided up into eight parts:

- Addition of numbers 0 - 10
- Subtraction of numbers of 0 - 10
- Addition of numbers 0 - 50
- Subtraction of numbers of 0 - 50
- Addition of numbers 0 - 100
- Subtraction of numbers of 0 - 100
- Mixed operations and money problems
- Timed tests

Some of the exercises are explained in more detail on page 5.

Answers are included at the back.

Good luck!

This book belongs to:

....

Table of Contents

Exercises Explained

Bullseye

The bullseye drill requires you to add the central number to the numbers on the inner ring or to subtract the central number from the numbers on the inner ring. Look for the sign in the centre of the bullseye!

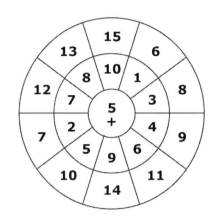

Match Ups

Match ups require you to complete the sum and draw a line to the correct answer on the opposite side.

a. 24 + 17 = ___	B•	•A = 38
b. 24 + 3 = ___	C•	•B = 41
c. 27 + 11 = ___	A•	•C = 27

Across Downs

Across Downs require you to add or subtract sums in a grid both horizontally and vertically.

This will provide numbers on the right-hand column and bottom row. Two final sums must then be written into the bottom-right cell.

6	+	2	+	0	=	8
+		+		+		+
3	+	4	+	3	=	10
+		+		+		+
7	+	2	+	6	=	15
=		=		=		=
16	+	8	+	9	=	33

Fact Families

Fact families are groups of facts containing the same numbers.

Two addition problems and two subtraction problems will be given each time.

93

23   70

23	+	70	=	93
70	+	23	=	93
93	-	23	=	70
93	-	70	=	23

5

Section 1: Adding Numbers 0-10

Name: _____   Date: _____

Class: _____   Teacher: _____

Adding Toucans

7 toucans

+

6 toucans

=

....toucans

1) $9 + 6$

2) $0 + 9$

3) $6 + 8$

4) $4 + 5$

5) $5 + 10$

6) $3 + 9$

7) $3 + 10$

8) $3 + 8$

9) $4 + 10$

10) $10 + 6$

11) $4 + 9$

12) $9 + 9$

13) $7 + 5$

14) $8 + 6$

15) $8 + 5$

16) $9 + 7$

17) $7 + 7$

18) $7 + 9$

19) $10 + 9$

20) $1 + 9$

Score: /20

Adding Numbers 0-10: Part 2

1) $\begin{array}{r} 7 \\ + 1 \\ \hline \end{array}$
2) $\begin{array}{r} 7 \\ + 9 \\ \hline \end{array}$
3) $\begin{array}{r} 10 \\ + 4 \\ \hline \end{array}$
4) $\begin{array}{r} 8 \\ + 7 \\ \hline \end{array}$
5) $\begin{array}{r} 10 \\ + 5 \\ \hline \end{array}$

6) $\begin{array}{r} 7 \\ + 0 \\ \hline \end{array}$
7) $\begin{array}{r} 6 \\ + 6 \\ \hline \end{array}$
8) $\begin{array}{r} 5 \\ + 6 \\ \hline \end{array}$
9) $\begin{array}{r} 7 \\ + 2 \\ \hline \end{array}$
10) $\begin{array}{r} 7 \\ + 7 \\ \hline \end{array}$

11) $\begin{array}{r} 6 \\ + 2 \\ \hline \end{array}$
12) $\begin{array}{r} 8 \\ + 10 \\ \hline \end{array}$
13) $\begin{array}{r} 9 \\ + 5 \\ \hline \end{array}$
14) $\begin{array}{r} 8 \\ + 9 \\ \hline \end{array}$
15) $\begin{array}{r} 7 \\ + 6 \\ \hline \end{array}$

16) $\begin{array}{r} 7 \\ + 8 \\ \hline \end{array}$
17) $\begin{array}{r} 9 \\ + 9 \\ \hline \end{array}$
18) $\begin{array}{r} 9 \\ + 8 \\ \hline \end{array}$
19) $\begin{array}{r} 6 \\ + 5 \\ \hline \end{array}$
20) $\begin{array}{r} 6 \\ + 9 \\ \hline \end{array}$

Score: _____ /20

Counting Spots!

Margot and Brian are twin giraffes. They both have 8 spots.

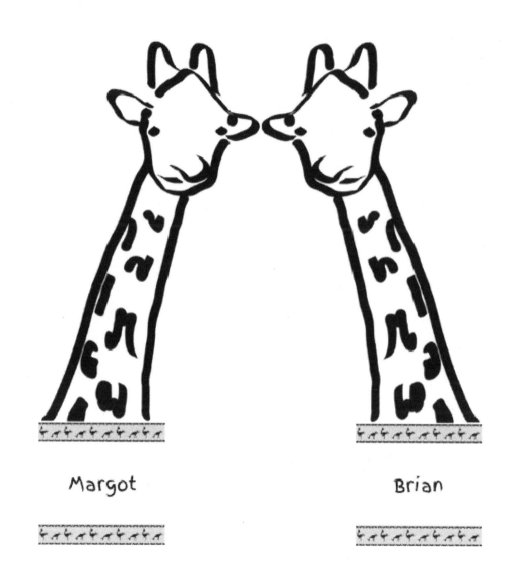

Margot

Brian

How many spots do they have altogether?

Adding Doubles 0-10

1) $\begin{array}{r} 2 \\ + 2 \\ \hline \end{array}$

2) $\begin{array}{r} 5 \\ + 5 \\ \hline \end{array}$

3) $\begin{array}{r} 4 \\ + 4 \\ \hline \end{array}$

4) $\begin{array}{r} 10 \\ + 10 \\ \hline \end{array}$

5) $\begin{array}{r} 6 \\ + 6 \\ \hline \end{array}$

6) $\begin{array}{r} 3 \\ + 3 \\ \hline \end{array}$

7) $\begin{array}{r} 1 \\ + 1 \\ \hline \end{array}$

8) $\begin{array}{r} 7 \\ + 7 \\ \hline \end{array}$

9) $\begin{array}{r} 8 \\ + 8 \\ \hline \end{array}$

10) $\begin{array}{r} 9 \\ + 9 \\ \hline \end{array}$

11) $\begin{array}{r} 6 \\ + 6 \\ \hline \end{array}$

12) $\begin{array}{r} 8 \\ + 8 \\ \hline \end{array}$

13) $\begin{array}{r} 2 \\ + 2 \\ \hline \end{array}$

14) $\begin{array}{r} 6 \\ + 6 \\ \hline \end{array}$

15) $\begin{array}{r} 6 \\ + 6 \\ \hline \end{array}$

Score: _____ /15

Adding 3 Single Digits

1)
```
   2
   6
+  8
____
```

2)
```
   6
   3
+  7
____
```

3)
```
   7
   7
+  8
____
```

4)
```
   3
   7
+  6
____
```

5)
```
   4
   9
+  7
____
```

6)
```
   2
   2
+  4
____
```

7)
```
   3
   3
+  8
____
```

8)
```
   7
   2
+  3
____
```

9)
```
   3
   1
+  6
____
```

Complete these addition boxes. The rows and columns add up to the boxes on the outside, don't forget the diagonals!

1)

				16
	3		9	
	7		21	
8	1	3	12	
20	11	11	15	

2)

				20
	8	9	25	
		9	23	
6		1	10	
23	16	19	14	

Score: _____ /11

Word Problems (Adding 0-10)

1) 2 red apples and 6 green apples are in the basket. How many apples are in the basket?

2) Charlotte has 9 coins but her friend Ellen gives her 8 more. How many coins does she have now?

3) Two flowers are growing. One has 9 petals and the other has 6. How many petals are there altogether?

Score: ___ /3

What can you add to the top number to make the bottom number?

1)
$$8$$
$$+ \underline{}$$
$$17$$

2)
$$6$$
$$+ \underline{}$$
$$14$$

3)
$$9$$
$$+ \underline{}$$
$$13$$

4)
$$7$$
$$+ \underline{}$$
$$19$$

5)
$$8$$
$$+ \underline{}$$
$$11$$

6)
$$7$$
$$+ \underline{}$$
$$13$$

7)
$$5$$
$$+ \underline{}$$
$$15$$

8)
$$8$$
$$+ \underline{}$$
$$17$$

9)
$$9$$
$$+ \underline{}$$
$$16$$

10)
$$9$$
$$+ \underline{}$$
$$13$$

11)
$$9$$
$$+ \underline{}$$
$$17$$

12)
$$9$$
$$+ \underline{}$$
$$15$$

13)
$$6$$
$$+ \underline{}$$
$$13$$

14)
$$8$$
$$+ \underline{}$$
$$10$$

15)
$$6$$
$$+ \underline{}$$
$$18$$

16)
$$6$$
$$+ \underline{}$$
$$14$$

17)
$$7$$
$$+ \underline{}$$
$$18$$

18)
$$9$$
$$+ \underline{}$$
$$16$$

19)
$$6$$
$$+ \underline{}$$
$$16$$

20)
$$8$$
$$+ \underline{}$$
$$19$$

Score: /20

Fill in the missing number!

1) $9 + 1 =$

2) $\_ + 3 = 12$

3) $\_ + 3 = 3$

4) $2 + 2 =$

5) $6 + 8 =$

6) $0 + 10 =$

7) $\_ + 1 = 5$

8) $4 + 5 =$

9) $\_ + 5 = 14$

10) $\_ + 1 = 2$

Score: _____ /10

1)

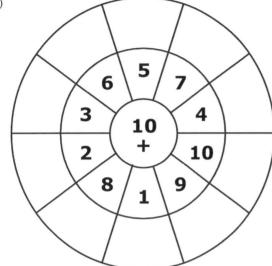

2)

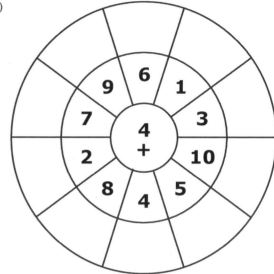

3)

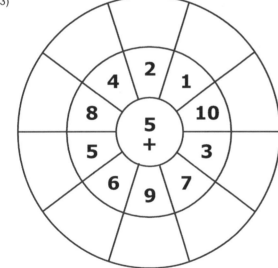

4)

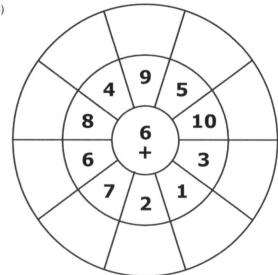

Score: /4

Match Ups: Addition 0-10

1)

a. 2 + 14 = _____ •   • D = 9

b. 6 + 17 = _____ •   • B = 15

c. 9 + 6 = _____ •   • C = 23

d. 2 + 15 = _____ •   • A = 7

e. 5 + 4 = _____ •   • F = 16

f. 3 + 8 = _____ •   • G = 17

g. 7 + 1 = _____ •   • E = 8

h. 4 + 3 = _____ •   • H = 11

Score: _18_

 3 tigers

+

 4 tigers

=

............................tigers

Section 2: Subtracting Numbers 0-10

Name: _____ Date: _____

Class: _____ Teacher: _____

Subtracting Zebras

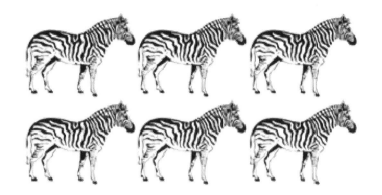

6 zebras

-

3 zebras

=

....zebras

Subtracting Numbers 0-10: Part 1

1) $$8 - 4$$

2) $$9 - 0$$

3) $$8 - 0$$

4) $$10 - 1$$

5) $$7 - 5$$

6) $$9 - 2$$

7) $$9 - 4$$

8) $$6 - 2$$

9) $$9 - 1$$

10) $$6 - 4$$

11) $$6 - 0$$

12) $$8 - 3$$

13) $$10 - 4$$

14) $$8 - 2$$

15) $$8 - 1$$

16) $$5 - 5$$

17) $$10 - 3$$

18) $$6 - 3$$

19) $$7 - 3$$

20) $$10 - 2$$

Score: /20

Subtracting Numbers 0-10: Part 2

1)　　7
　　- 7

2)　　7
　　- 4

3)　　9
　　- 6

4)　　6
　　- 3

5)　　6
　　- 5

6)　　9
　　- 5

7)　　5
　　- 1

8)　　10
　　- 5

9)　　8
　　- 5

10)　　7
　　- 2

11)　　7
　　- 5

12)　　6
　　- 2

13)　　6
　　- 4

14)　　6
　　- 1

15)　　9
　　- 3

16)　　8
　　- 0

17)　　6
　　- 6

18)　　10
　　- 9

19)　　8
　　- 2

20)　　7
　　- 3

Score: /20

Subtracting 3 Single Digits

1)
$$\begin{array}{r} 5 \\ -1 \\ -1 \\ \hline \end{array}$$

2)
$$\begin{array}{r} 4 \\ -3 \\ -1 \\ \hline \end{array}$$

3)
$$\begin{array}{r} 8 \\ -3 \\ -2 \\ \hline \end{array}$$

4)
$$\begin{array}{r} 9 \\ -1 \\ -2 \\ \hline \end{array}$$

5)
$$\begin{array}{r} 7 \\ -1 \\ -1 \\ \hline \end{array}$$

6)
$$\begin{array}{r} 7 \\ -3 \\ -3 \\ \hline \end{array}$$

7)
$$\begin{array}{r} 6 \\ -2 \\ -1 \\ \hline \end{array}$$

8)
$$\begin{array}{r} 6 \\ -3 \\ -3 \\ \hline \end{array}$$

9)
$$\begin{array}{r} 8 \\ -1 \\ -3 \\ \hline \end{array}$$

10)
$$\begin{array}{r} 7 \\ -2 \\ -1 \\ \hline \end{array}$$

11)
$$\begin{array}{r} 9 \\ -2 \\ -2 \\ \hline \end{array}$$

12)
$$\begin{array}{r} 4 \\ -2 \\ -1 \\ \hline \end{array}$$

13)
$$\begin{array}{r} 5 \\ -3 \\ -2 \\ \hline \end{array}$$

14)
$$\begin{array}{r} 4 \\ -1 \\ -3 \\ \hline \end{array}$$

15)
$$\begin{array}{r} 8 \\ -1 \\ -1 \\ \hline \end{array}$$

Score: _____ /15

1) **8 apples are in the basket. 4 are red and the rest are green. How many apples are green?**

2) **David has 2 marbles. Adam has 8 marbles. How many more marbles does Adam have than David?**

3) **10 peaches are in the basket. 4 peaches are taken out of the basket. How many peaches are in the basket now?**

Score: /3

What can you subtract from the top number to make the bottom number?

1)
$$\begin{array}{r} 8 \\ - \\ \hline 1 \end{array}$$

2)
$$\begin{array}{r} 5 \\ - \\ \hline 4 \end{array}$$

3)
$$\begin{array}{r} 2 \\ - \\ \hline 0 \end{array}$$

4)
$$\begin{array}{r} 1 \\ - \\ \hline 1 \end{array}$$

5)
$$\begin{array}{r} 9 \\ - \\ \hline 0 \end{array}$$

6)
$$\begin{array}{r} 10 \\ - \\ \hline 2 \end{array}$$

7)
$$\begin{array}{r} 6 \\ - \\ \hline 2 \end{array}$$

8)
$$\begin{array}{r} 7 \\ - \\ \hline 7 \end{array}$$

9)
$$\begin{array}{r} 7 \\ - \\ \hline 2 \end{array}$$

10)
$$\begin{array}{r} 3 \\ - \\ \hline 1 \end{array}$$

11)
$$\begin{array}{r} 4 \\ - \\ \hline 4 \end{array}$$

12)
$$\begin{array}{r} 0 \\ - \\ \hline 0 \end{array}$$

13)
$$\begin{array}{r} 3 \\ - \\ \hline 2 \end{array}$$

14)
$$\begin{array}{r} 2 \\ - \\ \hline 1 \end{array}$$

15)
$$\begin{array}{r} 4 \\ - \\ \hline 2 \end{array}$$

16)
$$\begin{array}{r} 4 \\ - \\ \hline 1 \end{array}$$

17)
$$\begin{array}{r} 8 \\ - \\ \hline 7 \end{array}$$

18)
$$\begin{array}{r} 1 \\ - \\ \hline 0 \end{array}$$

19)
$$\begin{array}{r} 8 \\ - \\ \hline 0 \end{array}$$

20)
$$\begin{array}{r} 9 \\ - \\ \hline 3 \end{array}$$

Score: ___ /20

Fill in the missing number!

1) $7 - 6 =$

2) $\_ - 8 = 1$

3) $9 - 6 =$

4) $\_ - 5 = 0$

5) $7 - 7 =$

6) $8 - 6 =$

7) $\_ - 5 = 4$

8) $9 - 7 =$

9) $9 - 9 =$

10) $10 - 6 =$

Score: _____ /10

Bullseye: Subtraction 0-10

1)

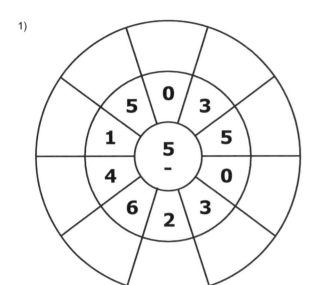

2)

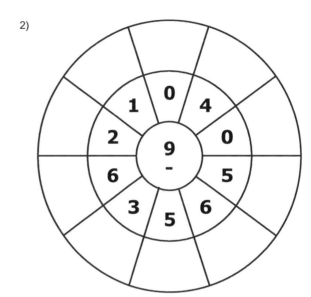

3)

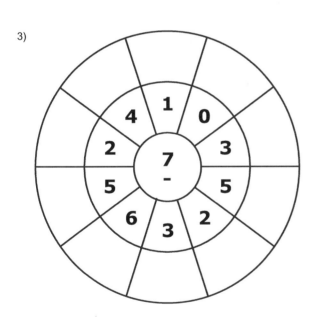

4)

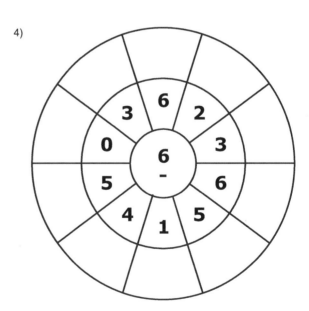

Score: /4

1)

a. $9 - 7 =$ _____ •   • E = 2

b. $3 - 2 =$ _____ •   • G = 1

c. $9 - 8 =$ _____ •   • F = 2

d. $7 - 6 =$ _____ •   • C = 0

e. $2 - 2 =$ _____ •   • A = 1

f. $1 - 1 =$ _____ •   • D = 5

g. $4 - 2 =$ _____ •   • B = 1

h. $10 - 5 =$ _____ •   • H = 0

Score: 18

Subtracting Elephants

5 elephants

−

2 elephants

=

....elephants

Section 3: Adding Numbers 0-50

Name: _____ Date: _____

Class: _____ Teacher: _____

Right or wrong?

This table has 4 incorrect sums. Highlight the wrong ones!

20 + 25 = 35	23 + 36 = 59	30 + 15 = 45
20 + 36 = 56	32 + 17 = 49	16 + 18 = 33
36 + 21 = 57	25 + 18 = 44	27 + 18 = 45
11 + 34 = 45	39 + 26 = 65	20 + 20 = 41

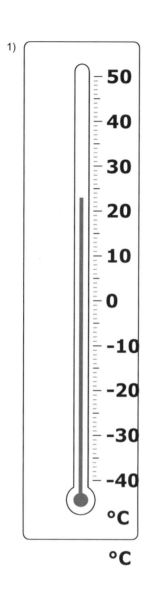

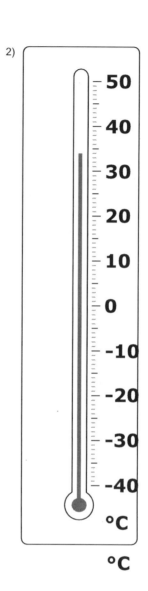

a) Write in the temperatures for both thermometers.

_____

b) Which thermometer is hotter? How much hotter is it?

_____

Adding Numbers 0-50: Part 1

1)
$$43 + 7$$

2)
$$35 + 10$$

3)
$$23 + 8$$

4)
$$40 + 9$$

5)
$$42 + 5$$

6)
$$28 + 3$$

7)
$$24 + 1$$

8)
$$38 + 4$$

9)
$$37 + 6$$

10)
$$42 + 7$$

11)
$$32 + 3$$

12)
$$44 + 6$$

13)
$$34 + 5$$

14)
$$21 + 4$$

15)
$$37 + 7$$

16)
$$39 + 3$$

17)
$$28 + 9$$

18)
$$22 + 7$$

19)
$$42 + 4$$

20)
$$23 + 6$$

Score: /20

Adding Numbers 0-50: Part 2

1)   6
  + 42

2)   2
  + 35

3)   9
  + 27

4)   5
  + 39

5)   6
  + 40

6)   5
  + 21

7)   7
  + 39

8)   3
  + 38

9)   9
  + 33

10)   3
  + 25

11)   9
  + 28

12)   2
  + 22

13)   1
  + 23

14)   3
  + 24

15)   6
  + 36

16)   0
  + 38

17)   1
  + 20

18)   2
  + 21

19)   0
  + 35

20)   5
  + 36

Score: /20

Adding Numbers 0-50: Part 3

1) 18
    $+ 30$

2) 21
    $+ 10$

3) 22
    $+ 20$

4) 11
    $+ 10$

5) 21
    $+ 30$

6) 31
    $+ 30$

7) 11
    $+ 20$

8) 24
    $+ 20$

9) 40
    $+ 10$

10) 29
    $+ 30$

11) 34
    $+ 10$

12) 31
    $+ 20$

13) 11
    $+ 30$

14) 15
    $+ 20$

15) 27
    $+ 10$

16) 37
    $+ 10$

17) 31
    $+ 30$

18) 38
    $+ 30$

19) 38
    $+ 40$

20) 17
    $+ 10$

Score: /20

Adding Numbers 0-50: Part 4

1)   35
   + 26

2)   26
   + 34

3)   27
   + 40

4)   32
   + 25

5)   39
   + 31

6)   42
   + 26

7)   33
   + 35

8)   45
   + 25

9)   33
   + 36

10)   31
   + 37

11)   36
   + 39

12)   35
   + 38

13)   26
   + 40

14)   47
   + 26

15)   26
   + 32

16)   30
   + 39

17)   27
   + 44

18)   26
   + 48

19)   40
   + 26

20)   33
   + 41

Score: /20

1)   28
   + 46

2)   41
   + 39

3)   40
   + 35

4)   38
   + 59

5)   53
   + 37

6)   46
   + 27

7)   27
   + 54

8)   33
   + 57

9)   42
   + 30

10)   29
   + 25

11)   44
   + 48

12)   49
   + 50

13)   50
   + 29

14)   42
   + 52

15)   33
   + 34

16)   39
   + 51

17)   39
   + 57

18)   42
   + 42

19)   37
   + 55

20)   44
   + 40

Score: _____ /20

Possible Sums: Addition

Add in numbers making possible pairs. Answers must be higher than 10.

$$\boxed{\_5} \; + \; \boxed{\_6} \; = \; \boxed{51}$$

15 + 36 | 25 + 26 | +

- -

$$\boxed{3\_} \; + \; \boxed{2\_} \; = \; \boxed{59}$$

30 + 29 | 31 + 28 | + | + | +

- -

+ | + | + | + | +

- -

$$\boxed{\_3} \; + \; \boxed{\_8} \; = \; \boxed{71}$$

13 + 58 | 23 + 48 | + | + | +

- -

1) 12
  + 12

2) 41
  + 41

3) 40
  + 40

4) 46
  + 46

5) 20
  + 20

6) 22
  + 22

7) 35
  + 35

8) 31
  + 31

9) 19
  + 19

10) 26
   + 26

11) 44
   + 44

12) 23
   + 23

13) 42
   + 42

14) 27
   + 27

15) 48
   + 48

Score: _____ /15

1)
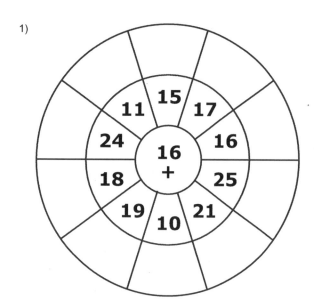

2)
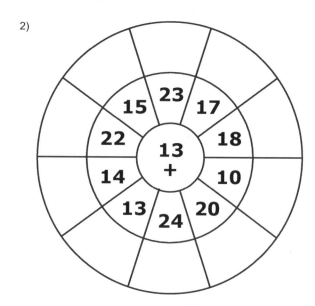

3)
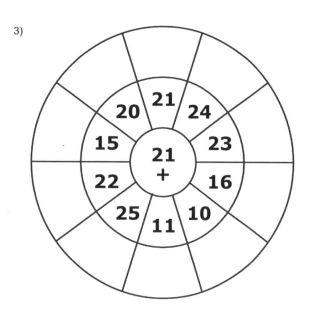

4)
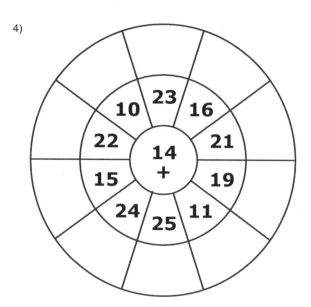

Score: ___/4___

Match Ups: Addition 0-50

1)

a. 2 + 6 = _____ •   • B = 48

b. 45 + 10 = _____ •   • A = 55

c. 21 + 10 = _____ •   • E = 60

d. 43 + 7 = _____ •   • H = 28

e. 3 + 4 = _____ •   • D = 50

f. 48 + 12 = _____ •   • G = 7

g. 37 + 11 = _____ •   • C = 8

h. 25 + 3 = _____ •   • F = 31

Score: 18

Across Downs: Addition 0-50

1)

14	+	**11**	+	**5**	=
+		+		+	+
15	+	**8**	+	**5**	=
+		+		+	+
12	+	**15**	+	**8**	=
=		=		=	=
	+		+		=

2)

11	+	**9**	+	**6**	=
+		+		+	+
13	+	**8**	+	**10**	=
+		+		+	+
13	+	**11**	+	**10**	=
=		=		=	=
	+		+		=

3)

11	+	**10**	+	**7**	=
+		+		+	+
5	+	**13**	+	**15**	=
+		+		+	+
10	+	**10**	+	**11**	=
=		=		=	=
	+		+		=

4)

8	+	**9**	+	**8**	=
+		+		+	+
5	+	**5**	+	**6**	=
+		+		+	+
12	+	**14**	+	**13**	=
=		=		=	=
	+		+		=

Score: _____ /4

Section 4: Subtracting Numbers 0-50

Name: _____ Date: _____

Class: _____ Teacher: _____

Right or Wrong?

This table has 4 incorrect sums. Highlight the wrong ones!

$41 - 26 = 15$	$43 - 32 = 11$	$41 - 38 = 3$
$41 - 27 = 14$	$43 - 32 = 1$	$40 - 26 = 15$
$41 - 25 = 16$	$35 - 21 = 14$	$27 - 24 = 6$
$33 - 22 = 13$	$47 - 36 = 11$	$22 - 11 = 11$

1)
$$\begin{array}{r} 14 \\ -\ 4 \\ \hline \end{array}$$

2)
$$\begin{array}{r} 20 \\ -\ 2 \\ \hline \end{array}$$

3)
$$\begin{array}{r} 28 \\ -\ 7 \\ \hline \end{array}$$

4)
$$\begin{array}{r} 23 \\ -\ 10 \\ \hline \end{array}$$

5)
$$\begin{array}{r} 40 \\ -\ 5 \\ \hline \end{array}$$

6)
$$\begin{array}{r} 19 \\ -\ 7 \\ \hline \end{array}$$

7)
$$\begin{array}{r} 39 \\ -\ 5 \\ \hline \end{array}$$

8)
$$\begin{array}{r} 15 \\ -\ 4 \\ \hline \end{array}$$

9)
$$\begin{array}{r} 38 \\ -\ 3 \\ \hline \end{array}$$

10)
$$\begin{array}{r} 46 \\ -\ 4 \\ \hline \end{array}$$

11)
$$\begin{array}{r} 31 \\ -\ 1 \\ \hline \end{array}$$

12)
$$\begin{array}{r} 13 \\ -\ 2 \\ \hline \end{array}$$

13)
$$\begin{array}{r} 22 \\ -\ 3 \\ \hline \end{array}$$

14)
$$\begin{array}{r} 13 \\ -\ 7 \\ \hline \end{array}$$

15)
$$\begin{array}{r} 20 \\ -\ 4 \\ \hline \end{array}$$

16)
$$\begin{array}{r} 13 \\ -\ 4 \\ \hline \end{array}$$

17)
$$\begin{array}{r} 47 \\ -\ 10 \\ \hline \end{array}$$

18)
$$\begin{array}{r} 21 \\ -\ 4 \\ \hline \end{array}$$

19)
$$\begin{array}{r} 34 \\ -\ 8 \\ \hline \end{array}$$

20)
$$\begin{array}{r} 44 \\ -\ 5 \\ \hline \end{array}$$

Score: _____ /20

Subtracting Numbers 0-50: Part 2

1) 29
   - 20

2) 46
   - 30

3) 14
   - 10

4) 31
   - 20

5) 39
   - 10

6) 44
   - 40

7) 48
   - 30

8) 35
   - 20

9) 14
   - 10

10) 29
    - 20

11) 42
    - 30

12) 46
    - 10

13) 44
    - 10

14) 45
    - 40

15) 40
    - 10

16) 35
    - 30

17) 41
    - 10

18) 42
    - 10

19) 48
    - 20

20) 43
    - 20

Score: _____ /20

Subtracting Numbers 0-50: Part 3

1) 33
  − 14

2) 40
  − 26

3) 42
  − 16

4) 30
  − 14

5) 42
  − 11

6) 13
  − 12

7) 38
  − 35

8) 20
  − 17

9) 29
  − 20

10) 38
  − 33

11) 29
  − 22

12) 16
  − 15

13) 44
  − 25

14) 11
  − 10

15) 33
  − 20

16) 37
  − 27

17) 18
  − 10

18) 24
  − 23

19) 34
  − 16

20) 41
  − 40

Score: _____ /20

Measuring Lines

1)

_____

2)

_____

3)

_____

4)

_____

5)

_____

6)

_____

7)

_____

8)

_____

9)

_____

10)

_____

a) Measure the lines. Which line is the longest, and which is the shortest?

_____

b) How much longer (in cm) is the longest line than the shortest?

_____

What can you subtract from the top number to make the bottom number?

1) 33
  − ___
    19

2) 48
  − ___
    29

3) 33
  − ___
    23

4) 13
  − ___
    3

5) 43
  − ___
    31

6) 32
  − ___
    19

7) 37
  − ___
    20

8) 47
  − ___
    5

9) 47
  − ___
    10

10) 24
  − ___
    2

11) 41
  − ___
    30

12) 37
  − ___
    3

13) 34
  − ___
    22

14) 22
  − ___
    12

15) 40
  − ___
    4

16) 38
  − ___
    4

17) 18
  − ___
    4

18) 21
  − ___
    2

19) 50
  − ___
    18

20) 15
  − ___
    3

Score: _____ /20

1)

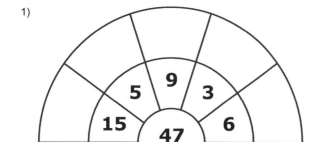

2)

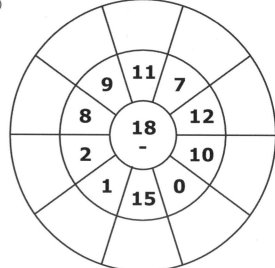

3)

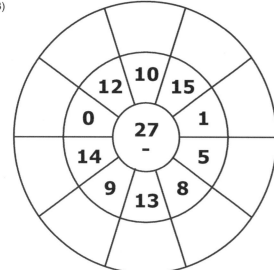

4)

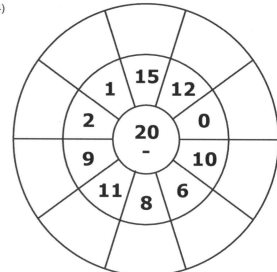

Score: _____ /4

Match Ups: Subtraction 0-50

1)

a. 23 - 1 = _____ •   • G = 19

b. 26 - 4 = _____ •   • D = 9

c. 26 - 7 = _____ •   • C = 18

d. 28 - 19 = _____ •   • F = 26

e. 43 - 15 = _____ •   • A = 22

f. 29 - 11 = _____ •   • E = 28

g. 31 - 5 = _____ •   • B = 2

h. 11 - 9 = _____ •   • H = 22

Score: 18

1)

42	-	11	-	20	=	
-		-		-		-
17	-	7	-	3	=	
-		-		-		-
12	-	3	-	8	=	
=		=		=		=
	-		-		=	

2)

52	-	26	-	21	=	
-		-		-		-
19	-	10	-	5	=	
-		-		-		-
18	-	7	-	10	=	
=		=		=		=
	-		-		=	

3)

46	-	17	-	21	=	
-		-		-		-
18	-	6	-	8	=	
-		-		-		-
11	-	1	-	10	=	
=		=		=		=
	-		-		=	

4)

40	-	9	-	18	=	
-		-		-		-
21	-	9	-	4	=	
-		-		-		-
12	-	0	-	7	=	
=		=		=		=
	-		-		=	

Score: _____ /4

Section 5: Adding Numbers 0-100

Name: _____   Date: _____

Class: _____   Teacher: _____

Number Lines

Complete these number lines.

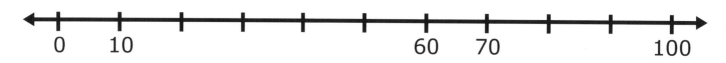

0   10   60   70   100

4   14   54   94

6   16   46   86   96

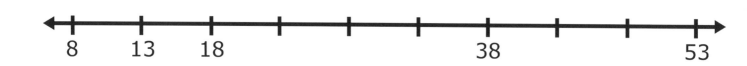

8   13   18   38   53

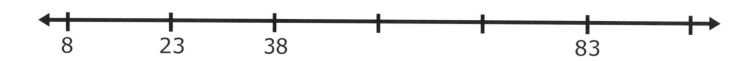

8   23   38   83

Part-whole Models

Fill in the missing numbers in the models below.

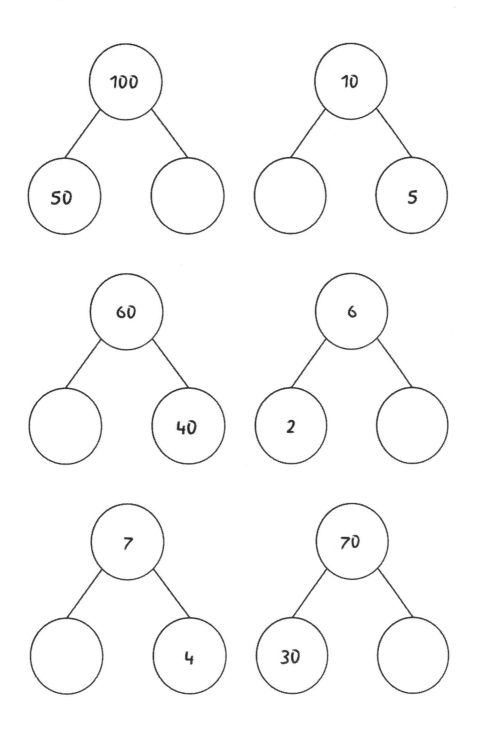

Adding Numbers 0-100: Part 1

Adding a single-digit number to a double-digit number.

1) 58 $+ \ 4$	2) 67 $+ \ 5$	3) 91 $+ \ 4$	4) 63 $+ \ 3$	5) 80 $+ \ 1$
6) 64 $+ \ 8$	7) 90 $+ 10$	8) 77 $+ \ 6$	9) 51 $+ \ 2$	10) 95 $+ \ 4$

Adding a double-digit number to a single-digit number.

1) 7 $+ 55$	2) 3 $+ 65$	3) 8 $+ 20$	4) 10 $+ 80$	5) 10 $+ 52$
6) 7 $+ 11$	7) 7 $+ 63$	8) 7 $+ 12$	9) 8 $+ 64$	10) 9 $+ 20$

Score: _____ /20

Solve these written questions.

a) What is 79 + 10?

_____

b) What does 8 tens + 3 ones equal?

_____

c) 10 + 5 + 10?

_____

d) What is the sum of 21 + 22?

_____

e) What is added to 75 to make 100?

_____

f) 5 tens + 9 ones?

_____

Adding Numbers 0-100: Part 3

Adding a multiple of ten to a double-digit number.

1)
$$61 + 10$$

2)
$$60 + 30$$

3)
$$53 + 30$$

4)
$$60 + 10$$

5)
$$65 + 30$$

6)
$$60 + 20$$

7)
$$59 + 10$$

8)
$$56 + 20$$

9)
$$67 + 30$$

10)
$$71 + 20$$

11)
$$50 + 30$$

12)
$$57 + 10$$

13)
$$72 + 20$$

14)
$$58 + 30$$

15)
$$64 + 10$$

16)
$$62 + 20$$

17)
$$66 + 10$$

18)
$$61 + 10$$

19)
$$74 + 20$$

20)
$$67 + 20$$

Score: _____ /20

Adding two double-digit numbers.

1)
$$56$$
$$+\ 44$$

2)
$$36$$
$$+\ 63$$

3)
$$58$$
$$+\ 38$$

4)
$$33$$
$$+\ 34$$

5)
$$43$$
$$+\ 42$$

6)
$$31$$
$$+\ 50$$

7)
$$31$$
$$+\ 38$$

8)
$$50$$
$$+\ 32$$

9)
$$33$$
$$+\ 44$$

10)
$$34$$
$$+\ 38$$

11)
$$56$$
$$+\ 39$$

12)
$$42$$
$$+\ 57$$

13)
$$37$$
$$+\ 41$$

14)
$$46$$
$$+\ 30$$

15)
$$38$$
$$+\ 31$$

16)
$$38$$
$$+\ 55$$

17)
$$49$$
$$+\ 47$$

18)
$$31$$
$$+\ 39$$

19)
$$47$$
$$+\ 43$$

20)
$$47$$
$$+\ 53$$

Score: /20

Number Bonds: Addition 0-100

What can you add to the top number to make the bottom number?

1)   64
   +____
     80

2)   55
   +____
     64

3)   68
   +____
     81

4)   61
   +____
     67

5)   57
   +____
     66

6)   50
   +____
     51

7)   75
   +____
     80

8)   54
   +____
     72

9)   58
   +____
     65

10)   72
   +____
     93

11)   66
   +____
     73

12)   51
   +____
     86

13)   52
   +____
     53

14)   57
   +____
     75

15)   79
   +____
     87

16)   75
   +____
     92

17)   67
   +____
     72

18)   57
   +____
     62

19)   70
   +____
     97

20)   51
   +____
     55

Score: _____ /20

Fill in the missing number!

1) **57 + 18 =**

2) **64 + 26 =**

3) **60 + 22 =**

4) **＿ + 10 = 70**

5) **62 + 31 =**

6) **＿ + 29 = 91**

7) **＿ + 49 = 99**

8) **＿ + 39 = 95**

9) **75 + 11 =**

10) **54 + 25 =**

Score: ＿＿＿ /10

Section 6: Subtracting Numbers 0-100

Name: _____ Date: _____

Class: _____ Teacher: _____

Complete the Models

Fill in the missing numbers and match the sums to their related facts.

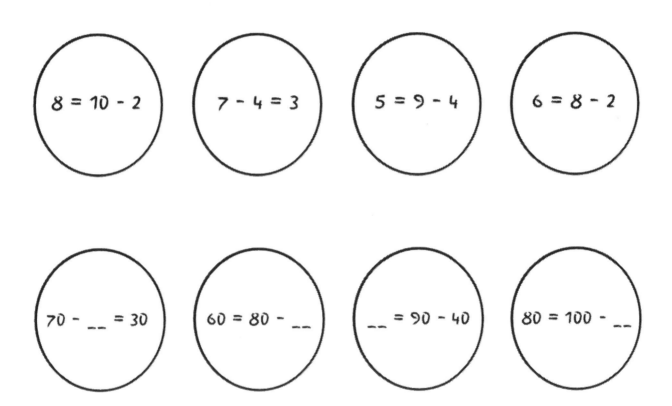

$8 = 10 - 2$

$7 - 4 = 3$

$5 = 9 - 4$

$6 = 8 - 2$

$70 - \_\_ = 30$

$60 = 80 - \_\_$

$\_\_ = 90 - 40$

$80 = 100 - \_\_$

Subtracting Numbers 0-100: Written Questions

Solve these written questions.

a) 84 - 10?

---

b) 50 minus 15?

---

c) What is 20 - 7 - 9?

---

d) What should be taken from 45 to make 36?

---

e) 81-28?

---

f) 6 tens minus 6 ones?

Solve these written questions.

a) What is half of 64?

b) 45 minus 29?

c) I own 21 pens, 3 red and the rest blue. How many are blue?

d) What do you subtract from 65 to make 46?

e) 71-30?

f) What is a quarter of 100?

Subtracting Numbers 0-100: Part 1

Subtracting a single-digit number from a double-digit number.

1) $63 - 3$

2) $23 - 1$

3) $84 - 9$

4) $18 - 8$

5) $65 - 7$

6) $99 - 1$

7) $73 - 8$

8) $98 - 2$

9) $72 - 2$

10) $28 - 7$

11) $97 - 6$

12) $91 - 9$

13) $83 - 9$

14) $64 - 5$

15) $29 - 2$

16) $50 - 2$

17) $75 - 8$

18) $22 - 6$

19) $70 - 4$

20) $96 - 7$

Score: /20

Subtracting a single-digit number from a double-digit number.

1)   60
   - 50
   ———

2)   91
   - 40
   ———

3)   83
   - 50
   ———

4)   87
   - 70
   ———

5)   91
   - 10
   ———

6)   71
   - 70
   ———

7)   81
   - 50
   ———

8)   86
   - 50
   ———

9)   92
   - 10
   ———

10)   80
   - 30
   ———

11)   99
   - 50
   ———

12)   83
   - 60
   ———

13)   78
   - 10
   ———

14)   62
   - 10
   ———

15)   77
   - 30
   ———

16)   66
   - 10
   ———

17)   81
   - 60
   ———

18)   57
   - 10
   ———

19)   75
   - 30
   ———

20)   86
   - 40
   ———

Score: _____ /20

Subtracting Numbers 0-100: Part 3

Subtracting a double-digit number from another double-digit number.

1) $\begin{array}{r} 70 \\ -\ 22 \\ \hline \end{array}$

2) $\begin{array}{r} 66 \\ -\ 48 \\ \hline \end{array}$

3) $\begin{array}{r} 63 \\ -\ 50 \\ \hline \end{array}$

4) $\begin{array}{r} 63 \\ -\ 43 \\ \hline \end{array}$

5) $\begin{array}{r} 61 \\ -\ 28 \\ \hline \end{array}$

6) $\begin{array}{r} 52 \\ -\ 44 \\ \hline \end{array}$

7) $\begin{array}{r} 52 \\ -\ 40 \\ \hline \end{array}$

8) $\begin{array}{r} 58 \\ -\ 42 \\ \hline \end{array}$

9) $\begin{array}{r} 71 \\ -\ 33 \\ \hline \end{array}$

10) $\begin{array}{r} 67 \\ -\ 37 \\ \hline \end{array}$

11) $\begin{array}{r} 56 \\ -\ 37 \\ \hline \end{array}$

12) $\begin{array}{r} 71 \\ -\ 22 \\ \hline \end{array}$

13) $\begin{array}{r} 51 \\ -\ 24 \\ \hline \end{array}$

14) $\begin{array}{r} 66 \\ -\ 25 \\ \hline \end{array}$

15) $\begin{array}{r} 51 \\ -\ 27 \\ \hline \end{array}$

16) $\begin{array}{r} 41 \\ -\ 32 \\ \hline \end{array}$

17) $\begin{array}{r} 63 \\ -\ 28 \\ \hline \end{array}$

18) $\begin{array}{r} 70 \\ -\ 38 \\ \hline \end{array}$

19) $\begin{array}{r} 45 \\ -\ 34 \\ \hline \end{array}$

20) $\begin{array}{r} 58 \\ -\ 24 \\ \hline \end{array}$

Score: /20

Number Bonds: Subtraction 0-100

What can you subtract from the top number to make the bottom number?

1) $\begin{array}{r} 64 \\ - \\ \hline 39 \end{array}$ 2) $\begin{array}{r} 87 \\ - \\ \hline 68 \end{array}$ 3) $\begin{array}{r} 61 \\ - \\ \hline 31 \end{array}$ 4) $\begin{array}{r} 48 \\ - \\ \hline 10 \end{array}$ 5) $\begin{array}{r} 91 \\ - \\ \hline 59 \end{array}$

6) $\begin{array}{r} 98 \\ - \\ \hline 70 \end{array}$ 7) $\begin{array}{r} 93 \\ - \\ \hline 76 \end{array}$ 8) $\begin{array}{r} 86 \\ - \\ \hline 67 \end{array}$ 9) $\begin{array}{r} 70 \\ - \\ \hline 49 \end{array}$ 10) $\begin{array}{r} 81 \\ - \\ \hline 73 \end{array}$

11) $\begin{array}{r} 53 \\ - \\ \hline 14 \end{array}$ 12) $\begin{array}{r} 73 \\ - \\ \hline 48 \end{array}$ 13) $\begin{array}{r} 52 \\ - \\ \hline 31 \end{array}$ 14) $\begin{array}{r} 77 \\ - \\ \hline 51 \end{array}$ 15) $\begin{array}{r} 58 \\ - \\ \hline 44 \end{array}$

16) $\begin{array}{r} 73 \\ - \\ \hline 69 \end{array}$ 17) $\begin{array}{r} 55 \\ - \\ \hline 53 \end{array}$ 18) $\begin{array}{r} 76 \\ - \\ \hline 60 \end{array}$ 19) $\begin{array}{r} 69 \\ - \\ \hline 45 \end{array}$ 20) $\begin{array}{r} 63 \\ - \\ \hline 30 \end{array}$

Score: _____ /20

Fill in the missing number!

1) $\_\_ - 12 = 85$

2) $81 - 12 = \_\_$

3) $\_\_ - 46 = 46$

4) $\_\_ - 9 = 90$

5) $\_\_ - 9 = 57$

6) $\_\_ - 33 = 41$

7) $\_\_ - 44 = 30$

8) $60 - 34 = \_\_$

9) $98 - 41 = \_\_$

10) $90 - 48 = \_\_$

Score: _____ /10

Section 7: Mixed Operations and Money

Name: _____   Date: _____

Class: _____   Teacher: _____

Fact Families 1

1)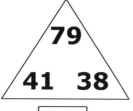

	+		=	
	+		=	
	-		=	
	-		=	

2)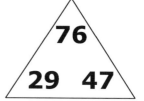

	+		=	
	+		=	
	-		=	
	-		=	

3)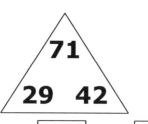

	+		=	
	+		=	
	-		=	
	-		=	

4)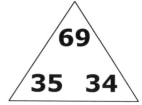

	+		=	
	+		=	
	-		=	
	-		=	

Score: /4

Fact Families 2

1)

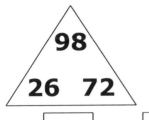

☐	+	☐	=	☐
☐	+	☐	=	☐
☐	-	☐	=	☐
☐	-	☐	=	☐

2)

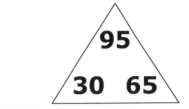

☐	+	☐	=	☐
☐	+	☐	=	☐
☐	-	☐	=	☐
☐	-	☐	=	☐

3)

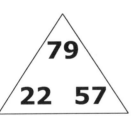

☐	+	☐	=	☐
☐	+	☐	=	☐
☐	-	☐	=	☐
☐	-	☐	=	☐

4)

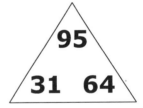

☐	+	☐	=	☐
☐	+	☐	=	☐
☐	-	☐	=	☐
☐	-	☐	=	☐

Score: _____ /4

1) $\begin{array}{r} 3 \\ -\ 0 \\ \hline \end{array}$
2) $\begin{array}{r} 6 \\ +\ 16 \\ \hline \end{array}$
3) $\begin{array}{r} 19 \\ -\ 4 \\ \hline \end{array}$
4) $\begin{array}{r} 16 \\ -\ 12 \\ \hline \end{array}$
5) $\begin{array}{r} 10 \\ -\ 1 \\ \hline \end{array}$

6) $\begin{array}{r} 8 \\ -\ 3 \\ \hline \end{array}$
7) $\begin{array}{r} 9 \\ +\ 7 \\ \hline \end{array}$
8) $\begin{array}{r} 8 \\ -\ 1 \\ \hline \end{array}$
9) $\begin{array}{r} 14 \\ +\ 5 \\ \hline \end{array}$
10) $\begin{array}{r} 19 \\ -\ 14 \\ \hline \end{array}$

11) $\begin{array}{r} 4 \\ +\ 4 \\ \hline \end{array}$
12) $\begin{array}{r} 3 \\ +\ 3 \\ \hline \end{array}$
13) $\begin{array}{r} 14 \\ -\ 2 \\ \hline \end{array}$
14) $\begin{array}{r} 19 \\ +\ 11 \\ \hline \end{array}$
15) $\begin{array}{r} 10 \\ -\ 5 \\ \hline \end{array}$

16) $\begin{array}{r} 7 \\ +\ 17 \\ \hline \end{array}$
17) $\begin{array}{r} 4 \\ +\ 9 \\ \hline \end{array}$
18) $\begin{array}{r} 1 \\ +\ 12 \\ \hline \end{array}$
19) $\begin{array}{r} 8 \\ +\ 13 \\ \hline \end{array}$
20) $\begin{array}{r} 11 \\ -\ 1 \\ \hline \end{array}$

Score: _____ /20

Adding and Subtracting: Part 2

1)
$$32 - 12$$

2)
$$46 - 26$$

3)
$$23 + 34$$

4)
$$34 - 10$$

5)
$$42 - 18$$

6)
$$48 - 25$$

7)
$$44 - 17$$

8)
$$34 - 7$$

9)
$$38 + 11$$

10)
$$38 + 19$$

11)
$$24 + 32$$

12)
$$31 + 32$$

13)
$$32 + 34$$

14)
$$23 + 29$$

15)
$$33 + 21$$

16)
$$45 - 18$$

17)
$$23 + 37$$

18)
$$46 - 25$$

19)
$$36 - 11$$

20)
$$31 + 34$$

Score: _____ /20

1) 25
    $+ 54$

2) 46
    $+ 13$

3) 57
    $- 34$

4) 50
    $- 12$

5) 38
    $- 12$

6) 57
    $- 22$

7) 72
    $- 47$

8) 38
    $- 23$

9) 31
    $+ 59$

10) 42
    $- 23$

11) 22
    $+ 48$

12) 43
    $- 18$

13) 52
    $- 12$

14) 70
    $+ 25$

15) 38
    $+ 25$

16) 52
    $- 35$

17) 35
    $+ 21$

18) 27
    $+ 16$

19) 54
    $+ 27$

20) 17
    $+ 37$

Score: _____ /20

1)
$$\begin{array}{r} 11 \\ -\ 10 \\ \hline \end{array}$$

2)
$$\begin{array}{r} 47 \\ +\ 47 \\ \hline \end{array}$$

3)
$$\begin{array}{r} 68 \\ +\ 31 \\ \hline \end{array}$$

4)
$$\begin{array}{r} 70 \\ +\ 23 \\ \hline \end{array}$$

5)
$$\begin{array}{r} 87 \\ -\ 82 \\ \hline \end{array}$$

6)
$$\begin{array}{r} 35 \\ -\ 29 \\ \hline \end{array}$$

7)
$$\begin{array}{r} 15 \\ -\ 10 \\ \hline \end{array}$$

8)
$$\begin{array}{r} 64 \\ +\ 30 \\ \hline \end{array}$$

9)
$$\begin{array}{r} 41 \\ +\ 57 \\ \hline \end{array}$$

10)
$$\begin{array}{r} 60 \\ -\ 52 \\ \hline \end{array}$$

11)
$$\begin{array}{r} 27 \\ +\ 61 \\ \hline \end{array}$$

12)
$$\begin{array}{r} 83 \\ -\ 77 \\ \hline \end{array}$$

13)
$$\begin{array}{r} 78 \\ +\ 19 \\ \hline \end{array}$$

14)
$$\begin{array}{r} 59 \\ -\ 53 \\ \hline \end{array}$$

15)
$$\begin{array}{r} 12 \\ -\ 10 \\ \hline \end{array}$$

16)
$$\begin{array}{r} 82 \\ +\ 15 \\ \hline \end{array}$$

17)
$$\begin{array}{r} 68 \\ -\ 66 \\ \hline \end{array}$$

18)
$$\begin{array}{r} 56 \\ -\ 52 \\ \hline \end{array}$$

19)
$$\begin{array}{r} 45 \\ +\ 36 \\ \hline \end{array}$$

20)
$$\begin{array}{r} 16 \\ +\ 62 \\ \hline \end{array}$$

Score: /20

Adding and Subtracting: Part 5

1) **79 - 17 =**

2) **17 + 24 =**

3) **93 - 83 =**

4) **26 - 24 =**

5) **32 - 20 =**

6) **20 - 11 =**

7) **39 - 12 =**

8) **32 + 12 =**

9) **56 - 27 =**

10) **94 - 71 =**

11) **61 + 22 =**

12) **18 + 35 =**

13) **11 + 75 =**

14) **12 + 17 =**

15) **48 - 34 =**

16) **41 + 35 =**

17) **40 - 17 =**

18) **20 + 42 =**

19) **43 + 37 =**

20) **28 + 61 =**

Score: /20

Matt has a 20p coin and a 50p coin. How much money does he have altogether?

Marianne has a 50p coin and a 5p coin. How much money does she have altogether?

John has a £5 note and Dan has a £10 note. How much money do they have altogether?

Victoria has a £20 note and a 50p coin. How much money does she have in total?

Joe has a £1 coin. He buys three chocolates, each costing 20p. How much money does he have left?

Penny has a 50p coin. She buys some sweets for 20p and a stamp for 10p. How much money does she have left?

Charlotte has a £20 note and a £10 note. She decides to spend half the total amount of money. How much is left?

What is 20p less than £5?

Shopping Problems

Solve these shopping problems. The answer bank shows all possible answers.

hot dog = £1.30	cola = £1.00
chips = £0.80	ice cream cone = £1.40
hamburger = £2.70	milk shake = £2.90
deluxe cheeseburger = £3.30	taco = £2.30

1) If Alice buys a taco and a cola, how much change will she get back from £10.00?

2) What is the total cost of a hamburger and a chips?

3) If Adam buys a chips and a deluxe cheeseburger, and if he had £10.00, how much money will he have left?

4) Steven wants to buy a hamburger, a chips, and a milk shake. How much will it cost him?

5) If David buys an ice cream cone, a milk shake, and a hot dog, how much change will he get back from £20.00?

6) If Michele buys a cola, and if she had £5.00, how much money will she have left?

7) What is the total cost of a hot dog and an ice cream cone?

8) If Dan wanted to buy a chips, how much would it cost him?

9) Billy wants to buy a hamburger. How much will he have to pay?

10) Anish purchases a hamburger, a cola, and a chips. How much money will he get back if he pays £10.00?

A. £6.70   B. £6.40   C. £2.70   D. £3.50   E. £4.00   F. £14.40
G. £0.80   H. £5.90   I. £2.70   J. £5.50

Score: _____ /10

Section 8: Timed Tests

Name: _____ Date: _____

Class: _____ Teacher: _____

Use this section to see how quickly and accurately you can do each page. See if you can beat your score and time each day!

Test 1: Addition 1

1) 6
  + 15

2) 2
  + 14

3) 13
  + 13

4) 8
  + 4

5) 9
  + 2

6) 2
  + 3

7) 4
  + 15

8) 2
  + 2

9) 1
  + 3

10) 6
  + 12

11) 4
  + 3

12) 5
  + 7

13) 15
  + 12

14) 8
  + 11

15) 13
  + 3

16) 12
  + 5

17) 6
  + 11

18) 3
  + 1

19) 12
  + 14

20) 10
  + 6

21) 14
  + 15

22) 0
  + 12

23) 3
  + 3

24) 2
  + 12

25) 9
  + 12

26) 9
  + 10

27) 12
  + 3

28) 10
  + 9

29) 15
  + 1

30) 5
  + 12

31) 6
  + 6

32) 6
  + 13

33) 12
  + 8

34) 8
  + 7

35) 14
  + 0

36) 6
  + 7

37) 4
  + 5

38) 7
  + 3

39) 9
  + 4

40) 9
  + 1

Test 2: Addition 2

1) 20
  + 15

2) 13
  + 25

3) 15
  + 15

4) 19
  + 15

5) 20
  + 20

6) 16
  + 21

7) 21
  + 12

8) 14
  + 17

9) 19
  + 19

10) 22
  + 16

11) 17
  + 13

12) 21
  + 17

13) 22
  + 14

14) 12
  + 16

15) 11
  + 22

16) 18
  + 24

17) 21
  + 22

18) 19
  + 21

19) 14
  + 23

20) 23
  + 23

21) 16
  + 16

22) 24
  + 17

23) 12
  + 19

24) 13
  + 13

25) 17
  + 11

26) 22
  + 13

27) 20
  + 24

28) 20
  + 17

29) 17
  + 14

30) 19
  + 16

31) 16
  + 11

32) 25
  + 23

33) 15
  + 14

34) 15
  + 24

35) 19
  + 13

36) 20
  + 13

37) 24
  + 16

38) 23
  + 14

39) 18
  + 17

40) 15
  + 21

Test 3: Addition 3

1) 15
   + 39

2) 38
   + 18

3) 13
   + 32

4) 21
   + 29

5) 26
   + 19

6) 36
   + 16

7) 33
   + 21

8) 30
   + 30

9) 20
   + 34

10) 31
    + 31

11) 19
    + 18

12) 13
    + 14

13) 25
    + 32

14) 30
    + 11

15) 27
    + 15

16) 13
    + 18

17) 34
    + 15

18) 20
    + 33

19) 17
    + 17

20) 20
    + 17

21) 31
    + 13

22) 20
    + 39

23) 37
    + 17

24) 38
    + 19

25) 12
    + 14

26) 33
    + 26

27) 18
    + 38

28) 17
    + 37

29) 36
    + 33

30) 33
    + 22

31) 12
    + 20

32) 13
    + 21

33) 15
    + 28

34) 21
    + 28

35) 11
    + 38

36) 23
    + 13

37) 24
    + 24

38) 40
    + 16

39) 38
    + 13

40) 17
    + 18

Test 4: Subtraction 1

1) 18
    $- 5$

2) 19
    $- 15$

3) 15
    $- 6$

4) 20
    $- 17$

5) 7
    $- 2$

6) 18
    $- 6$

7) 19
    $- 6$

8) 4
    $- 2$

9) 9
    $- 1$

10) 20
     $- 11$

11) 18
     $- 11$

12) 8
     $- 5$

13) 17
     $- 7$

14) 9
     $- 4$

15) 14
     $- 6$

16) 14
     $- 4$

17) 3
     $- 1$

18) 10
     $- 2$

19) 6
     $- 0$

20) 16
     $- 2$

21) 17
     $- 12$

22) 4
     $- 1$

23) 19
     $- 3$

24) 5
     $- 1$

25) 10
     $- 9$

26) 8
     $- 3$

27) 8
     $- 1$

28) 11
     $- 7$

29) 12
     $- 6$

30) 13
     $- 8$

31) 3
     $- 2$

32) 17
     $- 0$

33) 3
     $- 0$

34) 9
     $- 2$

35) 14
     $- 9$

36) 16
     $- 11$

37) 16
     $- 4$

38) 4
     $- 3$

39) 18
     $- 2$

40) 14
     $- 7$

Test 5: Subtraction 2

1)   27
   - 19

2)   20
   - 16

3)   22
   - 22

4)   16
   - 15

5)   20
   - 19

6)   24
   - 16

7)   39
   - 27

8)   36
   - 28

9)   37
   - 35

10)   32
   - 23

11)   31
   - 27

12)   21
   - 20

13)   35
   - 21

14)   19
   - 18

15)   17
   - 17

16)   36
   - 21

17)   35
   - 30

18)   18
   - 18

19)   26
   - 22

20)   33
   - 28

21)   35
   - 32

22)   17
   - 15

23)   18
   - 15

24)   33
   - 25

25)   36
   - 19

26)   31
   - 22

27)   39
   - 26

28)   18
   - 17

29)   39
   - 29

30)   34
   - 23

31)   22
   - 20

32)   32
   - 17

33)   27
   - 26

34)   27
   - 16

35)   34
   - 24

36)   23
   - 16

37)   39
   - 25

38)   33
   - 26

39)   19
   - 15

40)   26
   - 17

Test 6: Subtraction 3

1)   24
 − 24

2)   34
 − 32

3)   20
 − 20

4)   41
 − 33

5)   39
 − 35

6)   48
 − 23

7)   25
 − 23

8)   40
 − 23

9)   21
 − 21

10)   35
 − 29

11)   44
 − 26

12)   44
 − 25

13)   30
 − 20

14)   32
 − 25

15)   28
 − 20

16)   31
 − 30

17)   24
 − 22

18)   28
 − 21

19)   46
 − 41

20)   42
 − 22

21)   21
 − 20

22)   29
 − 28

23)   33
 − 28

24)   27
 − 25

25)   26
 − 21

26)   22
 − 22

27)   36
 − 34

28)   22
 − 21

29)   36
 − 21

30)   44
 − 23

31)   42
 − 42

32)   31
 − 29

33)   39
 − 38

34)   36
 − 31

35)   25
 − 20

36)   43
 − 32

37)   43
 − 30

38)   38
 − 27

39)   28
 − 23

40)   31
 − 22

Test 7: Addition and Subtraction 1

1) $\begin{array}{r} 15 \\ + 11 \\ \hline \end{array}$
2) $\begin{array}{r} 3 \\ + 2 \\ \hline \end{array}$
3) $\begin{array}{r} 4 \\ - 3 \\ \hline \end{array}$
4) $\begin{array}{r} 8 \\ + 7 \\ \hline \end{array}$
5) $\begin{array}{r} 7 \\ + 4 \\ \hline \end{array}$

6) $\begin{array}{r} 15 \\ - 7 \\ \hline \end{array}$
7) $\begin{array}{r} 13 \\ - 6 \\ \hline \end{array}$
8) $\begin{array}{r} 10 \\ - 1 \\ \hline \end{array}$
9) $\begin{array}{r} 17 \\ - 14 \\ \hline \end{array}$
10) $\begin{array}{r} 1 \\ + 7 \\ \hline \end{array}$

11) $\begin{array}{r} 14 \\ + 4 \\ \hline \end{array}$
12) $\begin{array}{r} 11 \\ + 19 \\ \hline \end{array}$
13) $\begin{array}{r} 8 \\ - 7 \\ \hline \end{array}$
14) $\begin{array}{r} 1 \\ + 8 \\ \hline \end{array}$
15) $\begin{array}{r} 7 \\ - 1 \\ \hline \end{array}$

16) $\begin{array}{r} 3 \\ + 1 \\ \hline \end{array}$
17) $\begin{array}{r} 11 \\ - 9 \\ \hline \end{array}$
18) $\begin{array}{r} 10 \\ + 15 \\ \hline \end{array}$
19) $\begin{array}{r} 18 \\ - 13 \\ \hline \end{array}$
20) $\begin{array}{r} 12 \\ - 7 \\ \hline \end{array}$

21) $\begin{array}{r} 4 \\ + 1 \\ \hline \end{array}$
22) $\begin{array}{r} 11 \\ - 8 \\ \hline \end{array}$
23) $\begin{array}{r} 3 \\ + 4 \\ \hline \end{array}$
24) $\begin{array}{r} 19 \\ - 18 \\ \hline \end{array}$
25) $\begin{array}{r} 15 \\ + 9 \\ \hline \end{array}$

26) $\begin{array}{r} 16 \\ - 12 \\ \hline \end{array}$
27) $\begin{array}{r} 16 \\ + 9 \\ \hline \end{array}$
28) $\begin{array}{r} 6 \\ - 5 \\ \hline \end{array}$
29) $\begin{array}{r} 10 \\ - 8 \\ \hline \end{array}$
30) $\begin{array}{r} 9 \\ - 4 \\ \hline \end{array}$

31) $\begin{array}{r} 13 \\ - 11 \\ \hline \end{array}$
32) $\begin{array}{r} 15 \\ + 16 \\ \hline \end{array}$
33) $\begin{array}{r} 12 \\ - 11 \\ \hline \end{array}$
34) $\begin{array}{r} 18 \\ - 10 \\ \hline \end{array}$
35) $\begin{array}{r} 13 \\ + 17 \\ \hline \end{array}$

36) $\begin{array}{r} 19 \\ + 4 \\ \hline \end{array}$
37) $\begin{array}{r} 12 \\ + 11 \\ \hline \end{array}$
38) $\begin{array}{r} 16 \\ + 14 \\ \hline \end{array}$
39) $\begin{array}{r} 15 \\ + 14 \\ \hline \end{array}$
40) $\begin{array}{r} 12 \\ - 3 \\ \hline \end{array}$

Test 8: Addition and Subtraction 2

1) $30 + 30$

2) $22 - 20$

3) $28 + 21$

4) $30 + 33$

5) $21 - 20$

6) $29 + 25$

7) $27 - 26$

8) $23 + 24$

9) $22 - 21$

10) $31 + 22$

11) $25 - 22$

12) $26 - 25$

13) $24 + 30$

14) $34 - 25$

15) $29 + 19$

16) $21 + 35$

17) $21 + 20$

18) $33 - 31$

19) $31 + 28$

20) $33 + 30$

21) $34 - 27$

22) $25 + 25$

23) $29 - 28$

24) $24 + 29$

25) $35 + 32$

26) $28 - 22$

27) $29 - 20$

28) $23 - 20$

29) $29 - 26$

30) $25 - 23$

31) $26 - 23$

32) $33 - 29$

33) $33 + 29$

34) $24 - 23$

35) $23 + 19$

36) $30 + 20$

37) $35 + 16$

38) $23 + 17$

39) $25 - 21$

40) $27 - 22$

Test 9: Addition and Subtraction 3

1)   59 − 53

2)   44 + 40

3)   40 + 41

4)   52 + 46

5)   66 − 37

6)   42 + 56

7)   18 − 10

8)   37 − 21

9)   45 + 43

10)   46 + 46

11)   41 + 46

12)   59 − 26

13)   47 + 40

14)   43 − 16

15)   42 + 41

16)   50 + 43

17)   51 + 48

18)   85 − 63

19)   80 − 30

20)   50 + 45

21)   66 − 12

22)   41 + 40

23)   44 − 19

24)   45 + 54

25)   55 − 29

26)   58 − 13

27)   44 − 36

28)   92 − 27

29)   71 − 12

30)   85 − 40

31)   40 + 46

32)   37 − 14

33)   42 + 47

34)   49 + 51

35)   51 + 43

36)   57 + 41

37)   41 + 55

38)   20 − 14

39)   16 − 12

40)   80 − 72

Test 10: Addition and Subtraction 4

1)   35
   + 57

2)   56
   + 32

3)   66
   − 63

4)   94
   − 85

5)   82
   − 76

6)   58
   + 40

7)   84
   − 76

8)   91
   − 89

9)   92
   − 86

10)   50
   + 42

11)   91
   − 84

12)   93
   − 87

13)   67
   − 61

14)   68
   − 63

15)   87
   − 82

16)   84
   − 82

17)   53
   + 36

18)   31
   + 42

19)   86
   − 82

20)   34
   + 64

21)   32
   + 38

22)   35
   + 43

23)   79
   − 74

24)   59
   + 38

25)   38
   + 50

26)   85
   − 82

27)   36
   + 45

28)   68
   + 32

29)   33
   + 54

30)   85
   − 83

31)   35
   + 54

32)   43
   + 52

33)   37
   + 56

34)   60
   + 32

35)   63
   + 31

36)   87
   − 81

37)   69
   − 65

38)   84
   − 80

39)   73
   − 69

40)   45
   + 34

ANSWERS

Section 1

Adding Toucans

13 toucans

Adding Numbers 0-10: Part 1

1) 9 + 6 = 15	2) 0 + 9 = 9	3) 6 + 8 = 14	4) 4 + 5 = 9	5) 5 + 10 = 15
6) 3 + 9 = 12	7) 3 + 10 = 13	8) 3 + 8 = 11	9) 4 + 10 = 14	10) 10 + 6 = 16
11) 4 + 9 = 13	12) 9 + 9 = 18	13) 7 + 5 = 12	14) 8 + 6 = 14	15) 8 + 5 = 13
16) 9 + 7 = 16	17) 7 + 7 = 14	18) 7 + 9 = 16	19) 10 + 9 = 19	20) 1 + 9 = 10

Adding Numbers 0-10: Part 2

1) 7 + 1 = 8	2) 7 + 9 = 16	3) 10 + 4 = 14	4) 8 + 7 = 15	5) 10 + 5 = 15
6) 7 + 0 = 7	7) 6 + 6 = 12	8) 5 + 6 = 11	9) 7 + 2 = 9	10) 7 + 7 = 14
11) 6 + 2 = 8	12) 8 + 10 = 18	13) 9 + 5 = 14	14) 8 + 9 = 17	15) 7 + 6 = 13
16) 7 + 8 = 15	17) 9 + 9 = 18	18) 9 + 8 = 17	19) 6 + 5 = 11	20) 6 + 9 = 15

Counting Spots! - 16 spots

Adding Doubles 0-10

1) 2 + 2 = 4	2) 5 + 5 = 10	3) 4 + 4 = 8
4) 10 + 10 = 20	5) 6 + 6 = 12	6) 3 + 3 = 6
7) 1 + 1 = 2	8) 7 + 7 = 14	9) 8 + 8 = 16
10) 9 + 9 = 18	11) 6 + 6 = 12	12) 8 + 8 = 16
13) 2 + 2 = 4	14) 6 + 6 = 12	15) 6 + 6 = 12

Adding 3 Single Digits

1) 2 + 6 + 8 = 16	2) 6 + 3 + 7 = 16	3) 7 + 7 + 8 = 22
4) 3 + 7 + 6 = 16	5) 4 + 9 + 7 = 20	6) 2 + 2 + 4 = 8
7) 3 + 3 + 8 = 14	8) 7 + 2 + 3 = 12	9) 3 + 1 + 6 = 10

Addition Boxes

1)

16		
5	3	1	9
7	7	7	21
8	1	3	12
20	11	11	15

2)

20		
8	8	9	25
9	5	9	23
6	3	1	10
23	16	19	14

Word Problems (Adding 0-10)

1) 8 2) 17 3) 15

Number Bonds: Addition 0-10

1) 8 + 9 = 17	2) 6 + 8 = 14	3) 9 + 4 = 13	4) 7 + 12 = 19	5) 8 + 3 = 11
6) 7 + 6 = 13	7) 5 + 10 = 15	8) 8 + 9 = 17	9) 9 + 7 = 16	10) 9 + 4 = 13
11) 9 + 8 = 17	12) 9 + 6 = 15	13) 6 + 7 = 13	14) 8 + 2 = 10	15) 6 + 12 = 18
16) 6 + 8 = 14	17) 7 + 11 = 18	18) 9 + 7 = 16	19) 6 + 10 = 16	20) 8 + 11 = 19

Number Bonds 2: Addition 0-10

1) 9 + 1 = 10	2) 9 + 3 = 12
3) 0 + 3 = 3	4) 2 + 2 = 4
5) 6 + 8 = 14	6) 0 + 10 = 10
7) 4 + 1 = 5	8) 4 + 5 = 9
9) 9 + 5 = 14	10) 1 + 1 = 2

Bullseye: Addition 0-10

1)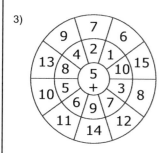

2)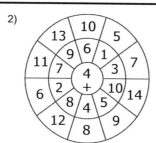

3)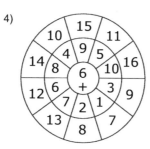

4)

Match Ups: Addition 0-10

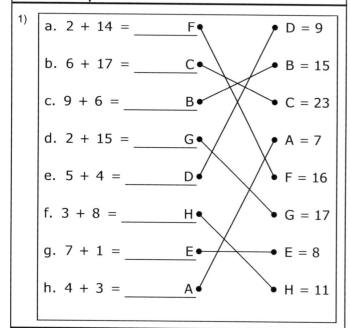

1)

a. 2 + 14 = _____ F

b. 6 + 17 = _____ C

c. 9 + 6 = _____ B

d. 2 + 15 = _____ G

e. 5 + 4 = _____ D

f. 3 + 8 = _____ H

g. 7 + 1 = _____ E

h. 4 + 3 = _____ A

D = 9
B = 15
C = 23
A = 7
F = 16
G = 17
E = 8
H = 11

Adding Tigers

7 tigers

Section 2

Subtracting Zebras

3 zebras

Subtracting Numbers 0-10: Part 1

1) 8 − 4 = 4	2) 9 − 0 = 9	3) 8 − 0 = 8	4) 10 − 1 = 9	5) 7 − 5 = 2
6) 9 − 2 = 7	7) 9 − 4 = 5	8) 6 − 2 = 4	9) 9 − 1 = 8	10) 6 − 4 = 2
11) 6 − 0 = 6	12) 8 − 3 = 5	13) 10 − 4 = 6	14) 8 − 2 = 6	15) 8 − 1 = 7
16) 5 − 5 = 0	17) 10 − 3 = 7	18) 6 − 3 = 3	19) 7 − 3 = 4	20) 10 − 2 = 8

Subtracting Numbers 0-10: Part 2

1) 7 − 7 = 0	2) 7 − 4 = 3	3) 9 − 6 = 3	4) 6 − 3 = 3	5) 6 − 5 = 1
6) 9 − 5 = 4	7) 5 − 1 = 4	8) 10 − 5 = 5	9) 8 − 5 = 3	10) 7 − 2 = 5
11) 7 − 5 = 2	12) 6 − 2 = 4	13) 6 − 4 = 2	14) 6 − 1 = 5	15) 9 − 3 = 6
16) 8 − 0 = 8	17) 6 − 6 = 0	18) 10 − 9 = 1	19) 8 − 2 = 6	20) 7 − 3 = 4

Subtracting 3 Single Digits

1) 5 − 1 − 1 = 3	2) 4 − 3 − 1 = 0	3) 8 − 3 − 2 = 3	4) 9 − 1 − 2 = 6	5) 7 − 1 − 1 = 5
6) 7 − 3 − 3 = 1	7) 6 − 2 − 1 = 3	8) 6 − 3 − 3 = 0	9) 8 − 1 − 3 = 4	10) 7 − 2 − 1 = 4
11) 9 − 2 − 2 = 5	12) 4 − 2 − 1 = 1	13) 5 − 3 − 2 = 0	14) 4 − 1 − 3 = 0	15) 8 − 1 − 1 = 6

96

Word Problems (Subtracting 0-10)

1) 4 2) 6 3) 6

Number Bonds: Subtraction 0-10

1) 8 − 7 = 1	2) 5 − 1 = 4	3) 2 − 2 = 0	4) 1 − 0 = 1	5) 9 − 9 = 0
6) 10 − 8 = 2	7) 6 − 4 = 2	8) 7 − 0 = 7	9) 7 − 5 = 2	10) 3 − 2 = 1
11) 4 − 0 = 4	12) 0 − 0 = 0	13) 3 − 1 = 2	14) 2 − 1 = 1	15) 4 − 2 = 2
16) 4 − 3 = 1	17) 8 − 1 = 7	18) 1 − 1 = 0	19) 8 − 8 = 0	20) 9 − 6 = 3

Number Bonds 2: Subtraction 0-10

1) $7 - 6 = 1$ 2) $9 - 8 = 1$
3) $9 - 6 = 3$ 4) $5 - 5 = 0$
5) $7 - 7 = 0$ 6) $8 - 6 = 2$
7) $9 - 5 = 4$ 8) $9 - 7 = 2$
9) $9 - 9 = 0$ 10) $10 - 6 = 4$

Bullseye: Subtraction 0-10

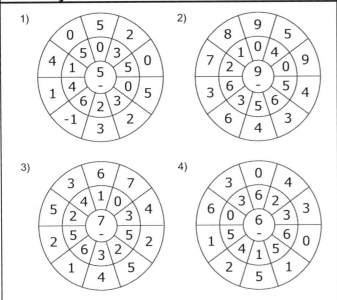

Match Ups: Subtraction 0-10

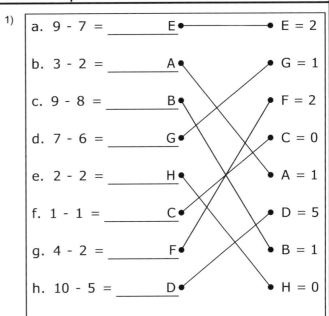

1)
a. $9 - 7 =$ ____ E
b. $3 - 2 =$ ____ A
c. $9 - 8 =$ ____ B
d. $7 - 6 =$ ____ G
e. $2 - 2 =$ ____ H
f. $1 - 1 =$ ____ C
g. $4 - 2 =$ ____ F
h. $10 - 5 =$ ____ D

E = 2
G = 1
F = 2
C = 0
A = 1
D = 5
B = 1
H = 0

Subtracting Elephants

3 elephants

Section 3

Right or wrong?

The following sums are incorrect:
20 + 25 = 35
16 + 18 = 33
25 + 18 = 44
20 + 20 = 41

Gauge the Heat!

a) 23°c and 34°c
b) Thermometer 2 is hotter by 11°c

Adding Numbers 0-50: Part 1

	1)	2)	3)	4)	5)
	43	35	23	40	42
	+ 7	+ 10	+ 8	+ 9	+ 5
	50	45	31	49	47
	6)	7)	8)	9)	10)
	28	24	38	37	42
	+ 3	+ 1	+ 4	+ 6	+ 7
	31	25	42	43	49
	11)	12)	13)	14)	15)
	32	44	34	21	37
	+ 3	+ 6	+ 5	+ 4	+ 7
	35	50	39	25	44
	16)	17)	18)	19)	20)
	39	28	22	42	23
	+ 3	+ 9	+ 7	+ 4	+ 6
	42	37	29	46	29

Adding Numbers 0-50: Part 2

	1)	2)	3)	4)	5)
	6	2	9	5	6
	+ 42	+ 35	+ 27	+ 39	+ 40
	48	37	36	44	46
	6)	7)	8)	9)	10)
	5	7	3	9	3
	+ 21	+ 39	+ 38	+ 33	+ 25
	26	46	41	42	28
	11)	12)	13)	14)	15)
	9	2	1	3	6
	+ 28	+ 22	+ 23	+ 24	+ 36
	37	24	24	27	42
	16)	17)	18)	19)	20)
	0	1	2	0	5
	+ 38	+ 20	+ 21	+ 35	+ 36
	38	21	23	35	41

Adding Numbers 0-50: Part 3

	1)	2)	3)	4)	5)
	18	21	22	11	21
	+ 30	+ 10	+ 20	+ 10	+ 30
	48	31	42	21	51
	6)	7)	8)	9)	10)
	31	11	24	40	29
	+ 30	+ 20	+ 20	+ 10	+ 30
	61	31	44	50	59
	11)	12)	13)	14)	15)
	34	31	11	15	27
	+ 10	+ 20	+ 30	+ 20	+ 10
	44	51	41	35	37
	16)	17)	18)	19)	20)
	37	31	38	38	17
	+ 10	+ 30	+ 30	+ 40	+ 10
	47	61	68	78	27

Adding Numbers 0-50: Part 4

	1)	2)	3)	4)	5)
	35	26	27	32	39
	+ 26	+ 34	+ 40	+ 25	+ 31
	61	60	67	57	70
	6)	7)	8)	9)	10)
	42	33	45	33	31
	+ 26	+ 35	+ 25	+ 36	+ 37
	68	68	70	69	68
	11)	12)	13)	14)	15)
	36	35	26	47	26
	+ 39	+ 38	+ 40	+ 26	+ 32
	75	73	66	73	58
	16)	17)	18)	19)	20)
	30	27	26	40	33
	+ 39	+ 44	+ 48	+ 26	+ 41
	69	71	74	66	74

Adding Numbers 0-50: Part 5

	1)	2)	3)	4)	5)
	28	41	40	38	53
	+ 46	+ 39	+ 35	+ 59	+ 37
	74	80	75	97	90
	6)	7)	8)	9)	10)
	46	27	33	42	29
	+ 27	+ 54	+ 57	+ 30	+ 25
	73	81	90	72	54
	11)	12)	13)	14)	15)
	44	49	50	42	33
	+ 48	+ 50	+ 29	+ 52	+ 34
	92	99	79	94	67
	16)	17)	18)	19)	20)
	39	39	42	37	44
	+ 51	+ 57	+ 42	+ 55	+ 40
	90	96	84	92	84

Possible Sums: Addition

Part 1:

15 + 36, 25 + 26, 35 + 16.

Part 2:

30 + 29, 31 + 28, 32 + 27, 33 + 26,

34 + 25, 35 + 24, 36 + 23, 37 + 22,

38 + 21, 39 + 20.

Part 3:

13 + 58, 23 + 48, 33 + 38, 43 + 28,

53 + 18.

Adding Doubles 0-50

1) 12 + 12 24	2) 41 + 41 82	3) 40 + 40 80
4) 46 + 46 92	5) 20 + 20 40	6) 22 + 22 44
7) 35 + 35 70	8) 31 + 31 62	9) 19 + 19 38
10) 26 + 26 52	11) 44 + 44 88	12) 23 + 23 46
13) 42 + 42 84	14) 27 + 27 54	15) 48 + 48 96

Bullseye: Addition 0-50

1)
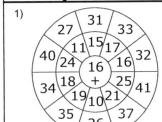

2)
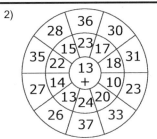

3)
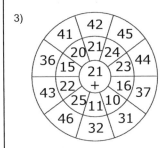

4)
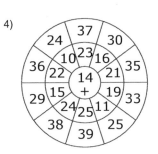

Match Ups: Addition 0-50

1)
a. 2 + 6 = _____ C

b. 45 + 10 = _____ A

c. 21 + 10 = _____ F

d. 43 + 7 = _____ D

e. 3 + 4 = _____ G

f. 48 + 12 = _____ E

g. 37 + 11 = _____ B

h. 25 + 3 = _____ H

B = 48
A = 55
E = 60
H = 28
D = 50
G = 7
C = 8
F = 31

Across Downs: Addition 0-50

1)

14	+	11	+	5	=	30
+		+		+		+
15	+	8	+	5	=	28
+		+		+		+
12	+	15	+	8	=	35
=		=		=		=
41	+	34	+	18	=	93

2)

11	+	9	+	6	=	26
+		+		+		+
13	+	8	+	10	=	31
+		+		+		+
13	+	11	+	10	=	34
=		=		=		=
37	+	28	+	26	=	91

3)

11	+	10	+	7	=	28
+		+		+		+
5	+	13	+	15	=	33
+		+		+		+
10	+	10	+	11	=	31
=		=		=		=
26	+	33	+	33	=	92

4)

8	+	9	+	8	=	25
+		+		+		+
5	+	5	+	6	=	16
+		+		+		+
12	+	14	+	13	=	39
=		=		=		=
25	+	28	+	27	=	80

Section 4

Right or wrong?

The following sums are incorrect:
43 - 32 = 1
40 - 26 = 15
27 - 24 = 6
33 - 22 = 13

Subtracting Numbers 0-50: Part 1

1) 14 - 4 10	2) 20 - 2 18	3) 28 - 7 21	4) 23 - 10 13	5) 40 - 5 35
6) 19 - 7 12	7) 39 - 5 34	8) 15 - 4 11	9) 38 - 3 35	10) 46 - 4 42
11) 31 - 1 30	12) 13 - 2 11	13) 22 - 3 19	14) 13 - 7 6	15) 20 - 4 16
16) 13 - 4 9	17) 47 - 10 37	18) 21 - 4 17	19) 34 - 8 26	20) 44 - 5 39

Subtracting Numbers 0-50: Part 2

1) 29 - 20 9	2) 46 - 30 16	3) 14 - 10 4	4) 31 - 20 11	5) 39 - 10 29
6) 44 - 40 4	7) 48 - 30 18	8) 35 - 20 15	9) 14 - 10 4	10) 29 - 20 9
11) 42 - 30 12	12) 46 - 10 36	13) 44 - 10 34	14) 45 - 40 5	15) 40 - 10 30
16) 35 - 30 5	17) 41 - 10 31	18) 42 - 10 32	19) 48 - 20 28	20) 43 - 20 23

Subtracting Numbers 0-50: Part 3

1) 33 - 14 19	2) 40 - 26 14	3) 42 - 16 26	4) 30 - 14 16	5) 42 - 11 31
6) 13 - 12 1	7) 38 - 35 3	8) 20 - 17 3	9) 29 - 20 9	10) 38 - 33 5
11) 29 - 22 7	12) 16 - 15 1	13) 44 - 25 19	14) 11 - 10 1	15) 33 - 20 13
16) 37 - 27 10	17) 18 - 10 8	18) 24 - 23 1	19) 34 - 16 18	20) 41 - 40 1

Measuring Lines

a) 10) is the longest and 4) is the shortest.

b) 14cm longer.

Number Bonds: Subtraction 0-50

1) 33 - 14 19	2) 48 - 19 29	3) 33 - 10 23	4) 13 - 10 3	5) 43 - 12 31
6) 32 - 13 19	7) 37 - 17 20	8) 47 - 42 5	9) 47 - 37 10	10) 24 - 22 2
11) 41 - 11 30	12) 37 - 34 3	13) 34 - 12 22	14) 22 - 10 12	15) 40 - 36 4
16) 38 - 34 4	17) 18 - 14 4	18) 21 - 19 2	19) 50 - 32 18	20) 15 - 12 3

Bullseye: Subtraction 0-50

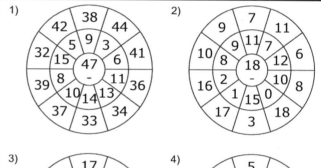

Match Ups: Subtraction 0-50

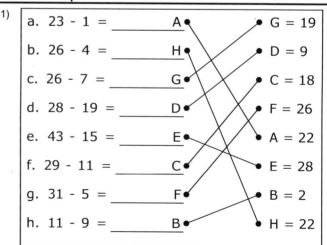

a. 23 - 1 = ____ A
b. 26 - 4 = ____ H
c. 26 - 7 = ____ G
d. 28 - 19 = ____ D
e. 43 - 15 = ____ E
f. 29 - 11 = ____ C
g. 31 - 5 = ____ F
h. 11 - 9 = ____ B

G = 19
D = 9
C = 18
F = 26
A = 22
E = 28
B = 2
H = 22

Across Downs: Subtraction 0-50

1)

42	-	11	-	20	=	11
-		-		-		-
17	-	7	-	3	=	7
-		-		-		-
12	-	3	-	8	=	1
=		=		=		=
13	-	1	-	9	=	3

2)

52	-	26	-	21	=	5
-		-		-		-
19	-	10	-	5	=	4
-		-		-		-
18	-	7	-	10	=	1
=		=		=		=
15	-	9	-	6	=	0

3)

46	-	17	-	21	=	8
-		-		-		-
18	-	6	-	8	=	4
-		-		-		-
11	-	1	-	10	=	0
=		=		=		=
17	-	10	-	3	=	4

4)

40	-	9	-	18	=	13
-		-		-		-
21	-	9	-	4	=	8
-		-		-		-
12	-	0	-	7	=	5
=		=		=		=
7	-	0	-	7	=	0

Section 5

Number Lines

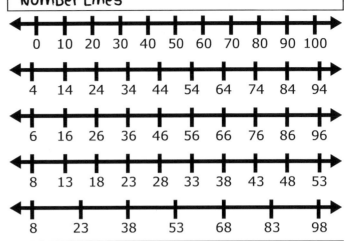

Part-Whole Models

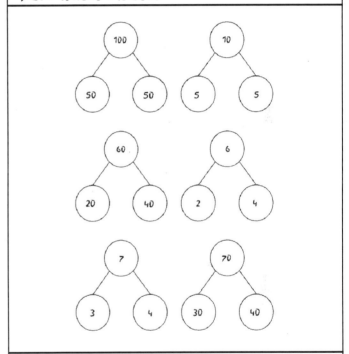

Adding Numbers 0-100: Part 1

1) $\begin{array}{r} 58 \\ + \ 4 \\ \hline 62 \end{array}$	2) $\begin{array}{r} 67 \\ + \ 5 \\ \hline 72 \end{array}$	3) $\begin{array}{r} 91 \\ + \ 4 \\ \hline 95 \end{array}$	4) $\begin{array}{r} 63 \\ + \ 3 \\ \hline 66 \end{array}$	5) $\begin{array}{r} 80 \\ + \ 1 \\ \hline 81 \end{array}$
6) $\begin{array}{r} 64 \\ + \ 8 \\ \hline 72 \end{array}$	7) $\begin{array}{r} 90 \\ + 10 \\ \hline 100 \end{array}$	8) $\begin{array}{r} 77 \\ + \ 6 \\ \hline 83 \end{array}$	9) $\begin{array}{r} 51 \\ + \ 2 \\ \hline 53 \end{array}$	10) $\begin{array}{r} 95 \\ + \ 4 \\ \hline 99 \end{array}$

1) $\begin{array}{r} 7 \\ + 55 \\ \hline 62 \end{array}$	2) $\begin{array}{r} 3 \\ + 65 \\ \hline 68 \end{array}$	3) $\begin{array}{r} 8 \\ + 20 \\ \hline 28 \end{array}$	4) $\begin{array}{r} 10 \\ + 80 \\ \hline 90 \end{array}$	5) $\begin{array}{r} 10 \\ + 52 \\ \hline 62 \end{array}$
6) $\begin{array}{r} 7 \\ + 11 \\ \hline 18 \end{array}$	7) $\begin{array}{r} 7 \\ + 63 \\ \hline 70 \end{array}$	8) $\begin{array}{r} 7 \\ + 12 \\ \hline 19 \end{array}$	9) $\begin{array}{r} 8 \\ + 64 \\ \hline 72 \end{array}$	10) $\begin{array}{r} 9 \\ + 20 \\ \hline 29 \end{array}$

Adding Numbers 0-100: Part 2 (Written Questions)

a) 89

b) 83

c) 25

d) 43

e) 25

f) 59

Adding Numbers 0-100: Part 3

1)	2)	3)	4)	5)
61 + 10 71	60 + 30 90	53 + 30 83	60 + 10 70	65 + 30 95
6)	7)	8)	9)	10)
60 + 20 80	59 + 10 69	56 + 20 76	67 + 30 97	71 + 20 91
11)	12)	13)	14)	15)
50 + 30 80	57 + 10 67	72 + 20 92	58 + 30 88	64 + 10 74
16)	17)	18)	19)	20)
62 + 20 82	66 + 10 76	61 + 10 71	74 + 20 94	67 + 20 87

Adding Numbers 0-100: Part 4

1)	2)	3)	4)	5)
56 + 44 100	36 + 63 99	58 + 38 96	33 + 34 67	43 + 42 85
6)	7)	8)	9)	10)
31 + 50 81	31 + 38 69	50 + 32 82	33 + 44 77	34 + 38 72
11)	12)	13)	14)	15)
56 + 39 95	42 + 57 99	37 + 41 78	46 + 30 76	38 + 31 69
16)	17)	18)	19)	20)
38 + 55 93	49 + 47 96	31 + 39 70	47 + 43 90	47 + 53 100

Number Bonds: Addition 0-100

1)	2)	3)	4)	5)
64 + 16 80	55 + 9 64	68 + 13 81	61 + 6 67	57 + 9 66
6)	7)	8)	9)	10)
50 + 1 51	75 + 5 80	54 + 18 72	58 + 7 65	72 + 21 93
11)	12)	13)	14)	15)
66 + 7 73	51 + 35 86	52 + 1 53	57 + 18 75	79 + 8 87
16)	17)	18)	19)	20)
75 + 17 92	67 + 5 72	57 + 5 62	70 + 27 97	51 + 4 55

Number Bonds 2: Addition 0-100

1) $57 + 18 = 75$

2) $64 + 26 = 90$

3) $60 + 22 = 82$

4) $60 + 10 = 70$

5) $62 + 31 = 93$

6) $62 + 29 = 91$

7) $50 + 49 = 99$

8) $56 + 39 = 95$

9) $75 + 11 = 86$

10) $54 + 25 = 79$

Section 6

Complete the Models

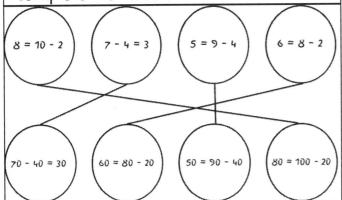

Circles top row: 8 = 10 - 2 | 7 - 4 = 3 | 5 = 9 - 4 | 6 = 8 - 2

Circles bottom row: 70 - 40 = 30 | 60 = 80 - 20 | 50 = 90 - 40 | 80 = 100 - 20

Subtracting Numbers 0-100: Written Questions

a) 74

b) 35

c) 4

d) 9

e) 53

f) 54

Subtracting Numbers 0-100: Written Questions 2

a) 32

b) 16

c) 18

d) 19

e) 41

f) 25

Subtracting Numbers 0-100: Part 1

1) 63 − 3 = 60	2) 23 − 1 = 22	3) 84 − 9 = 75	4) 18 − 8 = 10	5) 65 − 7 = 58
6) 99 − 1 = 98	7) 73 − 8 = 65	8) 98 − 2 = 96	9) 72 − 2 = 70	10) 28 − 7 = 21
11) 97 − 6 = 91	12) 91 − 9 = 82	13) 83 − 9 = 74	14) 64 − 5 = 59	15) 29 − 2 = 27
16) 50 − 2 = 48	17) 75 − 8 = 67	18) 22 − 6 = 16	19) 70 − 4 = 66	20) 96 − 7 = 89

Subtracting Numbers 0-100: Part 2

1) 60 − 50 = 10	2) 91 − 40 = 51	3) 83 − 50 = 33	4) 87 − 70 = 17	5) 91 − 10 = 81
6) 71 − 70 = 1	7) 81 − 50 = 31	8) 86 − 50 = 36	9) 92 − 10 = 82	10) 80 − 30 = 50
11) 99 − 50 = 49	12) 83 − 60 = 23	13) 78 − 10 = 68	14) 62 − 10 = 52	15) 77 − 30 = 47
16) 66 − 10 = 56	17) 81 − 60 = 21	18) 57 − 10 = 47	19) 75 − 30 = 45	20) 86 − 40 = 46

Subtracting Numbers 0-100: Part 3

1) 70 − 22 = 48	2) 66 − 48 = 18	3) 63 − 50 = 13	4) 63 − 43 = 20	5) 61 − 28 = 33
6) 52 − 44 = 8	7) 52 − 40 = 12	8) 58 − 42 = 16	9) 71 − 33 = 38	10) 67 − 37 = 30
11) 56 − 37 = 19	12) 71 − 22 = 49	13) 51 − 24 = 27	14) 66 − 25 = 41	15) 51 − 27 = 24
16) 41 − 32 = 9	17) 63 − 28 = 35	18) 70 − 38 = 32	19) 45 − 34 = 11	20) 58 − 24 = 34

Number Bonds: Subtraction 0-100

1) 64 − 25 = 39	2) 87 − 19 = 68	3) 61 − 30 = 31	4) 48 − 38 = 10	5) 91 − 32 = 59
6) 98 − 28 = 70	7) 93 − 17 = 76	8) 86 − 19 = 67	9) 70 − 21 = 49	10) 81 − 8 = 73
11) 53 − 39 = 14	12) 73 − 25 = 48	13) 52 − 21 = 31	14) 77 − 26 = 51	15) 58 − 14 = 44
16) 73 − 4 = 69	17) 55 − 2 = 53	18) 76 − 16 = 60	19) 69 − 24 = 45	20) 63 − 33 = 30

Number Bonds 2: Subtraction 0-100

1) 97 - 12 = 85

2) 81 - 12 = 69

3) 92 - 46 = 46

4) 99 - 9 = 90

5) 66 - 9 = 57

6) 74 - 33 = 41

7) 74 - 44 = 30

8) 60 - 34 = 26

9) 98 - 41 = 57

10) 90 - 48 = 42

Section 7

Fact Families 1

1)

79
41 38

41	+	38	=	79
38	+	41	=	79
79	-	41	=	38
79	-	38	=	41

2)

76
29 47

29	+	47	=	76
47	+	29	=	76
76	-	29	=	47
76	-	47	=	29

3)

71
29 42

29	+	42	=	71
42	+	29	=	71
71	-	29	=	42
71	-	42	=	29

4)

69
35 34

35	+	34	=	69
34	+	35	=	69
69	-	35	=	34
69	-	34	=	35

Fact Families 2

1)

98
26 72

26	+	72	=	98
72	+	26	=	98
98	-	26	=	72
98	-	72	=	26

2)

95
30 65

30	+	65	=	95
65	+	30	=	95
95	-	30	=	65
95	-	65	=	30

3)

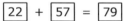

79
22 57

22	+	57	=	79
57	+	22	=	79
79	-	22	=	57
79	-	57	=	22

4)

95
31 64

31	+	64	=	95
64	+	31	=	95
95	-	31	=	64
95	-	64	=	31

Adding and Subtracting: Part 1

1)	2)	3)	4)	5)
3 - 0 3	6 + 16 22	19 - 4 15	16 - 12 4	10 - 1 9
6) 8 - 3 5	7) 9 + 7 16	8) 8 - 1 7	9) 14 + 5 19	10) 19 - 14 5
11) 4 + 4 8	12) 3 + 3 6	13) 14 - 2 12	14) 19 + 11 30	15) 10 - 5 5
16) 7 + 17 24	17) 4 + 9 13	18) 1 + 12 13	19) 8 + 13 21	20) 11 - 1 10

Adding and Subtracting: Part 2

1)	2)	3)	4)	5)
32 - 12 20	46 - 26 20	23 + 34 57	34 - 10 24	42 - 18 24
6) 48 - 25 23	7) 44 - 17 27	8) 34 - 7 27	9) 38 + 11 49	10) 38 + 19 57
11) 24 + 32 56	12) 31 + 32 63	13) 32 + 34 66	14) 23 + 29 52	15) 33 + 21 54
16) 45 - 18 27	17) 23 + 37 60	18) 46 - 25 21	19) 36 - 11 25	20) 31 + 34 65

Adding and Subtracting: Part 3

1)	2)	3)	4)	5)
25 + 54 79	46 + 13 59	57 - 34 23	50 - 12 38	38 - 12 26
6) 57 - 22 35	7) 72 - 47 25	8) 38 - 23 15	9) 31 + 59 90	10) 42 - 23 19
11) 22 + 48 70	12) 43 - 18 25	13) 52 - 12 40	14) 70 + 25 95	15) 38 + 25 63
16) 52 - 35 17	17) 35 + 21 56	18) 27 + 16 43	19) 54 + 27 81	20) 17 + 37 54

Adding and Subtracting: Part 4

1)	2)	3)	4)	5)
11 - 10 1	47 + 47 94	68 + 31 99	70 + 23 93	87 - 82 5
6) 35 - 29 6	7) 15 - 10 5	8) 64 + 30 94	9) 41 + 57 98	10) 60 - 52 8
11) 27 + 61 88	12) 83 - 77 6	13) 78 + 19 97	14) 59 - 53 6	15) 12 - 10 2
16) 82 + 15 97	17) 68 - 66 2	18) 56 - 52 4	19) 45 + 36 81	20) 16 + 62 78

Adding and Subtracting: Part 5

1) 79 - 17 = 62
2) 17 + 24 = 41
3) 93 - 83 = 10
4) 26 - 24 = 2
5) 32 - 20 = 12
6) 20 - 11 = 9
7) 39 - 12 = 27
8) 32 + 12 = 44
9) 56 - 27 = 29
10) 94 - 71 = 23
11) 61 + 22 = 83
12) 18 + 35 = 53
13) 11 + 75 = 86
14) 12 + 17 = 29
15) 48 - 34 = 14
16) 41 + 35 = 76
17) 40 - 17 = 23
18) 20 + 42 = 62
19) 43 + 37 = 80
20) 28 + 61 = 89

Adding Money

Matt has 70p

Marianne has 55p.

Adding Money 2

John and Dan have £15.

Victoria has £20.50.

Subtracting Money

Joe has 40p.

Penny has 20p.

Subtracting Money 2

Charlotte has £15.

£5.00 - 20p = £4.80.

Shopping Problems

1) £6.70

2) £3.50

3) £5.90

4) £6.40

5) £14.40

6) £4.00

7) £2.70

8) £0.80

9) £2.70

10) £5.50

Section 8

Test 1

#		#		#		#		#	
1)	6 + 15 = 21	2)	2 + 14 = 16	3)	13 + 13 = 26	4)	8 + 4 = 12	5)	9 + 2 = 11
6)	2 + 3 = 5	7)	4 + 15 = 19	8)	2 + 2 = 4	9)	1 + 3 = 4	10)	6 + 12 = 18
11)	4 + 3 = 7	12)	5 + 7 = 12	13)	15 + 12 = 27	14)	8 + 11 = 19	15)	13 + 3 = 16
16)	12 + 5 = 17	17)	6 + 11 = 17	18)	3 + 1 = 4	19)	12 + 14 = 26	20)	10 + 6 = 16
21)	14 + 15 = 29	22)	0 + 12 = 12	23)	3 + 3 = 6	24)	2 + 12 = 14	25)	9 + 12 = 21
26)	9 + 10 = 19	27)	12 + 3 = 15	28)	10 + 9 = 19	29)	15 + 1 = 16	30)	5 + 12 = 17
31)	6 + 6 = 12	32)	6 + 13 = 19	33)	12 + 8 = 20	34)	8 + 7 = 15	35)	14 + 0 = 14
36)	6 + 7 = 13	37)	4 + 5 = 9	38)	7 + 3 = 10	39)	9 + 4 = 13	40)	9 + 1 = 10

Test 2

#		#		#		#		#	
1)	20 + 15 = 35	2)	13 + 25 = 38	3)	15 + 15 = 30	4)	19 + 15 = 34	5)	20 + 20 = 40
6)	16 + 21 = 37	7)	21 + 12 = 33	8)	14 + 17 = 31	9)	19 + 19 = 38	10)	22 + 16 = 38
11)	17 + 13 = 30	12)	21 + 17 = 38	13)	22 + 14 = 36	14)	12 + 16 = 28	15)	11 + 22 = 33
16)	18 + 24 = 42	17)	21 + 22 = 43	18)	19 + 21 = 40	19)	14 + 23 = 37	20)	23 + 23 = 46
21)	16 + 16 = 32	22)	24 + 17 = 41	23)	12 + 19 = 31	24)	13 + 13 = 26	25)	17 + 11 = 28
26)	22 + 13 = 35	27)	20 + 24 = 44	28)	20 + 17 = 37	29)	17 + 14 = 31	30)	19 + 16 = 35
31)	16 + 11 = 27	32)	25 + 23 = 48	33)	15 + 14 = 29	34)	15 + 24 = 39	35)	19 + 13 = 32
36)	20 + 13 = 33	37)	24 + 16 = 40	38)	23 + 14 = 37	39)	18 + 17 = 35	40)	15 + 21 = 36

Test 3

1) 15 + 39 = 54
2) 38 + 18 = 56
3) 13 + 32 = 45
4) 21 + 29 = 50
5) 26 + 19 = 45
6) 36 + 16 = 52
7) 33 + 21 = 54
8) 30 + 30 = 60
9) 20 + 34 = 54
10) 31 + 31 = 62
11) 19 + 18 = 37
12) 13 + 14 = 27
13) 25 + 32 = 57
14) 30 + 11 = 41
15) 27 + 15 = 42
16) 13 + 18 = 31
17) 34 + 15 = 49
18) 20 + 33 = 53
19) 17 + 17 = 34
20) 20 + 17 = 37
21) 31 + 13 = 44
22) 20 + 39 = 59
23) 37 + 17 = 54
24) 38 + 19 = 57
25) 12 + 14 = 26
26) 33 + 26 = 59
27) 18 + 38 = 56
28) 17 + 37 = 54
29) 36 + 33 = 69
30) 33 + 22 = 55
31) 12 + 20 = 32
32) 13 + 21 = 34
33) 15 + 28 = 43
34) 21 + 28 = 49
35) 11 + 38 = 49
36) 23 + 13 = 36
37) 24 + 24 = 48
38) 40 + 16 = 56
39) 38 + 13 = 51
40) 17 + 18 = 35

Test 4

1) 18 − 5 = 13
2) 19 − 15 = 4
3) 15 − 6 = 9
4) 20 − 17 = 3
5) 7 − 2 = 5
6) 18 − 6 = 12
7) 19 − 6 = 13
8) 4 − 2 = 2
9) 9 − 1 = 8
10) 20 − 11 = 9
11) 18 − 11 = 7
12) 8 − 5 = 3
13) 17 − 7 = 10
14) 9 − 4 = 5
15) 14 − 6 = 8
16) 14 − 4 = 10
17) 3 − 1 = 2
18) 10 − 2 = 8
19) 6 − 0 = 6
20) 16 − 2 = 14
21) 17 − 12 = 5
22) 4 − 1 = 3
23) 19 − 3 = 16
24) 5 − 1 = 4
25) 10 − 9 = 1
26) 8 − 3 = 5
27) 8 − 1 = 7
28) 11 − 7 = 4
29) 12 − 6 = 6
30) 13 − 8 = 5
31) 3 − 2 = 1
32) 17 − 0 = 17
33) 3 − 0 = 3
34) 9 − 2 = 7
35) 14 − 9 = 5
36) 16 − 11 = 5
37) 16 − 4 = 12
38) 4 − 3 = 1
39) 18 − 2 = 16
40) 14 − 7 = 7

Test 5

1) 27 − 19 = 8
2) 20 − 16 = 4
3) 22 − 22 = 0
4) 16 − 15 = 1
5) 20 − 19 = 1
6) 24 − 16 = 8
7) 39 − 27 = 12
8) 36 − 28 = 8
9) 37 − 35 = 2
10) 32 − 23 = 9
11) 31 − 27 = 4
12) 21 − 20 = 1
13) 35 − 21 = 14
14) 19 − 18 = 1
15) 17 − 17 = 0
16) 36 − 21 = 15
17) 35 − 30 = 5
18) 18 − 18 = 0
19) 26 − 22 = 4
20) 33 − 28 = 5
21) 35 − 32 = 3
22) 17 − 15 = 2
23) 18 − 15 = 3
24) 33 − 25 = 8
25) 36 − 19 = 17
26) 31 − 22 = 9
27) 39 − 26 = 13
28) 18 − 17 = 1
29) 39 − 29 = 10
30) 34 − 23 = 11
31) 22 − 20 = 2
32) 32 − 17 = 15
33) 27 − 26 = 1
34) 27 − 16 = 11
35) 34 − 24 = 10
36) 23 − 16 = 7
37) 39 − 25 = 14
38) 33 − 26 = 7
39) 19 − 15 = 4
40) 26 − 17 = 9

Test 6

1) 24 − 24 = 0
2) 34 − 32 = 2
3) 20 − 20 = 0
4) 41 − 33 = 8
5) 39 − 35 = 4
6) 48 − 23 = 25
7) 25 − 23 = 2
8) 40 − 23 = 17
9) 21 − 21 = 0
10) 35 − 29 = 6
11) 44 − 26 = 18
12) 44 − 25 = 19
13) 30 − 20 = 10
14) 32 − 25 = 7
15) 28 − 20 = 8
16) 31 − 30 = 1
17) 24 − 22 = 2
18) 28 − 21 = 7
19) 46 − 41 = 5
20) 42 − 22 = 20
21) 21 − 20 = 1
22) 29 − 28 = 1
23) 33 − 28 = 5
24) 27 − 25 = 2
25) 26 − 21 = 5
26) 22 − 22 = 0
27) 36 − 34 = 2
28) 22 − 21 = 1
29) 36 − 21 = 15
30) 44 − 23 = 21
31) 42 − 42 = 0
32) 31 − 29 = 2
33) 39 − 38 = 1
34) 36 − 31 = 5
35) 25 − 20 = 5
36) 43 − 32 = 11
37) 43 − 30 = 13
38) 38 − 27 = 11
39) 28 − 23 = 5
40) 31 − 22 = 9

Test 7

1) 15 + 11 = 26
2) 3 + 2 = 5
3) 4 − 3 = 1
4) 8 + 7 = 15
5) 7 + 4 = 11
6) 15 − 7 = 8
7) 13 − 6 = 7
8) 10 − 1 = 9
9) 17 − 14 = 3
10) 1 + 7 = 8
11) 14 + 4 = 18
12) 11 + 19 = 30
13) 8 − 7 = 1
14) 1 + 8 = 9
15) 7 − 1 = 6
16) 3 + 1 = 4
17) 11 − 9 = 2
18) 10 + 15 = 25
19) 18 − 13 = 5
20) 12 − 7 = 5
21) 4 + 1 = 5
22) 11 − 8 = 3
23) 3 + 4 = 7
24) 19 − 18 = 1
25) 15 + 9 = 24
26) 16 − 12 = 4
27) 16 + 9 = 25
28) 6 − 5 = 1
29) 10 − 8 = 2
30) 9 − 4 = 5
31) 13 − 11 = 2
32) 15 + 16 = 31
33) 12 − 11 = 1
34) 18 − 10 = 8
35) 13 + 17 = 30
36) 19 + 4 = 23
37) 12 + 11 = 23
38) 16 + 14 = 30
39) 15 + 14 = 29
40) 12 − 3 = 9

Test 8

1) 30 + 30 = 60
2) 22 − 20 = 2
3) 28 + 21 = 49
4) 30 + 33 = 63
5) 21 − 20 = 1
6) 29 + 25 = 54
7) 27 − 26 = 1
8) 23 + 24 = 47
9) 22 − 21 = 1
10) 31 + 22 = 53
11) 25 − 22 = 3
12) 26 − 25 = 1
13) 24 + 30 = 54
14) 34 − 25 = 9
15) 29 + 19 = 48
16) 21 + 35 = 56
17) 21 + 20 = 41
18) 33 − 31 = 2
19) 31 + 28 = 59
20) 33 + 30 = 63
21) 34 − 27 = 7
22) 25 + 25 = 50
23) 29 − 28 = 1
24) 24 + 29 = 53
25) 35 + 32 = 67
26) 28 − 22 = 6
27) 29 − 20 = 9
28) 23 − 20 = 3
29) 29 − 26 = 3
30) 25 − 23 = 2
31) 26 − 23 = 3
32) 33 − 29 = 4
33) 33 + 29 = 62
34) 24 − 23 = 1
35) 23 + 19 = 42
36) 30 + 20 = 50
37) 35 + 16 = 51
38) 23 + 17 = 40
39) 25 − 21 = 4
40) 27 − 22 = 5

Test 9

1) 59 − 53 = 6
2) 44 + 40 = 84
3) 40 + 41 = 81
4) 52 + 46 = 98
5) 66 − 37 = 29
6) 42 + 56 = 98
7) 18 − 10 = 8
8) 37 − 21 = 16
9) 45 + 43 = 88
10) 46 + 46 = 92
11) 41 + 46 = 87
12) 59 − 26 = 33
13) 47 + 40 = 87
14) 43 − 16 = 27
15) 42 + 41 = 83
16) 50 + 43 = 93
17) 51 + 48 = 99
18) 85 − 63 = 22
19) 80 − 30 = 50
20) 50 + 45 = 95
21) 66 − 12 = 54
22) 41 + 40 = 81
23) 44 − 19 = 25
24) 45 + 54 = 99
25) 55 − 29 = 26
26) 58 − 13 = 45
27) 44 − 36 = 8
28) 92 − 27 = 65
29) 71 − 12 = 59
30) 85 − 40 = 45
31) 40 + 46 = 86
32) 37 − 14 = 23
33) 42 + 47 = 89
34) 49 + 51 = 100
35) 51 + 43 = 94
36) 57 + 41 = 98
37) 41 + 55 = 96
38) 20 − 14 = 6
39) 16 − 12 = 4
40) 80 − 72 = 8

Test 10

1) 35 + 57 = 92
2) 56 + 32 = 88
3) 66 − 63 = 3
4) 94 − 85 = 9
5) 82 − 76 = 6
6) 58 + 40 = 98
7) 84 − 76 = 8
8) 91 − 89 = 2
9) 92 − 86 = 6
10) 50 + 42 = 92
11) 91 − 84 = 7
12) 93 − 87 = 6
13) 67 − 61 = 6
14) 68 − 63 = 5
15) 87 − 82 = 5
16) 84 − 82 = 2
17) 53 + 36 = 89
18) 31 + 42 = 73
19) 86 − 82 = 4
20) 34 + 64 = 98
21) 32 + 38 = 70
22) 35 + 43 = 78
23) 79 − 74 = 5
24) 59 + 38 = 97
25) 38 + 50 = 88
26) 85 − 82 = 3
27) 36 + 45 = 81
28) 68 + 32 = 100
29) 33 + 54 = 87
30) 85 − 83 = 2
31) 35 + 54 = 89
32) 43 + 52 = 95
33) 37 + 56 = 93
34) 60 + 32 = 92
35) 63 + 31 = 94
36) 87 − 81 = 6
37) 69 − 65 = 4
38) 84 − 80 = 4
39) 73 − 69 = 4
40) 45 + 34 = 79

Printed in Great Britain
by Amazon